AF877384

To Paul
God Bless the
Schultz' Family

Ken

POULTRY
OF THE WORLD

By Loyl Stromberg

First published 1996 by Silvio Mattacchione & Co.

1251 Scugog Line 8, RR #1,
Port Perry, Ontario, Canada L9L 1B2
Telephone: 1-905-985-3555
Fax: 1-905-985-4005
Orders only 1-800-779-4163

Text and photography
© 1996 Silvio Mattacchione & Co.,

All rights reserved. No part of this publication may be reproduced or transmitted in any form or by any means, electronic or mechanical including photocopying, recording or any information storage or retrieval system, without the prior permission in writing from the publishers.

Canadian Cataloguing in Publication Data
Stromberg, Loyl, 1916-
Stromberg's poultry of the world

Includes bibliographical references.
ISBN 1-895270-16-2

1. Poultry breeds. 2. Poultry. I. Title.
II. Title: Poultry of the world.

SF487.S77 1996 636.5 C95-932872-6

Distribution for the United States and Canada
Silvio Mattacchione & Co
1251 Scugog Line 8, RR #1,
Port Perry, Ontario, Canada L9L 1B2
Telephone: 1-905-985-3555
Fax: 1-905-985-4005
Orders only 1-800-779-4163

Design
Peter A. Graziano Limited
1923 Spruce Hill Road
Pickering, Ontario, Canada
L1V 1S6
Phone/Fax (905) 837-7718

Printed and bound in Canada by
D.W. Friesen & Sons Ltd.

DEDICATION

This book is dedicated
to my parents,
Ernst and Josephine Stromberg
and to my late wife, Evelyn.
In 1921, on a farm near Otho, Iowa,
my parents were pioneers in the
commercial chick hatchery industry.
As the enterprise grew,
they found they needed better mail and
transportation facilities so they left the
farm in 1924 and moved to
Fort Dodge, Iowa.
Between 1924 and 1965
the Stromberg Hatchery
grew into one of the
largest chick hatcheries
in the United States.
It was the experience I gained
from years of working in the
hatchery business
and then on my own,
that provided me
with the background and desire
to write this book.

Loyl and Evelyn Stromberg

Table of Contents

ACKNOWLEDGMENTS

Poultry of the World is the result of extensive research and would not have been possible without the cooperation of many people from around the world. It is difficult to give credit and recognition to all those who so generously shared both research material and knowledge. If I have missed anyone, please forgive me and advise, and we will give appropriate credit in future reprints. It is with grateful and sincere appreciation that I introduce my friends.

Chang Ching Wang	*China*
Hans L. Schippers	*Holland*
Anna C. Banning Vogelpoel	*Holland*
Noboru Suzuki	*Japan*
Dr. Willard Hollander	*USA*
Dr. William Longenecker	*USA*

I would also like to thank the following people;

William Plant	*Australia*
Dr. Edmund Hoffman	*Canada*
Professor Xiangpin Qiu	*China*
Dr. Mohammed Said Samy	*Egypt*
Tom Bartlett	*England*
Will Burdett,	*England*
Reverend Ray Trudigan	*England*
P. R. Gupta	*India*
P. D. Selvaraj	*India*
Yoshijiro Kajima	*Japan*
Professor Valdimar Fisinin	*Russia*

Stained glass window in Stromberg Library

Loyl Stromberg

ABOUT THE AUTHOR

Loyl Stromberg is known and respected world-wide. He began his lifelong career in the poultry industry in 1921 at the age of seven. He is the author of 14 books on poultry and poultry related subjects. His son, Loy, and daughter-in-law, Janet, operate The Stromberg Publishing Company, Pine River, Minnesota, which he began in 1969.

Stromberg was instrumental in the formation of the National Poultry Museum which is a multi-million dollar venture in Bonners Springs, Kansas. It is an integral part of the Agricultural Museum and Farmer's Hall of Fame and scheduled to open in August, 1998. The first exhibits opened to the public in November, 1994. The Agricultural Museum/Farmer's Hall of Fame attracts over 25,000 visitors each year.

The objective of the National Poultry Museum is to preserve poultry history, present facts and to educate the public regarding poultry progress and values. Their goal is to present a good public image of the world of commercial poultry and the egg industry and at the same time to remind the public of the pleasure of poultry exhibition as a hobby.

In 1971, Loyl Stromberg re-organized the SPPA (The Society for the Preservation of Poultry Antiquities). Their foremost service is the breeders source directory.

Presented with permission of the Imperial Household of Japan

PHOTOGRAPHY

Gary Hart, Australia
Lance J. Ruting, Australia
John Tarren, England
Rudi Proll, Germany
Josef Wolters, Germany
Mitsuyoshi Tatematsu, Japan
Hans Schippers, Netherlands
E.A. Zhigailov, Russia
Dr. Jeffrey Black, USA
Dr. William Harris, USA
Richard Holmes, USA

ARTWORK

Janice Brent, USA
Mary Ellen Johnson, USA

Additional thanks to;
Charles G. Olentine, Jr., Vice-President and Publisher of Watt Publishing and Poultry Group for permission to quote from Poultry Tribune.

The Board of Chamberlains of the Imperial Palace of Tokyo for permission to use the photograph *Gunkei-Zu* and or *Gunkei-No-Zu* by Ito.

Patricia Bascom, Publications Officer, Olasgu Museum and Art Galleries, for permission to reproduce the Roman mosaic of *The Cockerel* from the Burrell Collection.

Tom Fay, Director of Marketing, Out of the West Publishing, for permission to use his postcard *High Stepper,* and to Arti Grafiche, Stampe-d'arte.

Frank Oppel, Marketing Director, Book Sales, Inc., Secacus, NJ, for permission to quote items listed from their book, *Why Do We Say It?*

Christopher Groves, Marketing Director, Guinness Books, for permission to quote facts from the *Guinness Book of Records.*

"My Hen"
by
Geraldine Knight
English, Contemporary
The Sladmore Gallery

Painting
by
Mary Ellen Johnson

Anna Bianconi, Gruppo Giornalistico Edagricole, for her photographs.

Roxanna Ring Rosenstiels, Fine Art Publishers, for permission to reproduce *To the Pasture New* by Sir James Guthrie.

Stitching Vriend Van Het Mauritshuis, for permission to reproduce *The Poultry Yard* by Jan Steen.

Ruth M. Winston, National Geographic Magazine, for permission to reproduce artwork appearing in the April, 1927 issue of National Geographic.

VAAP, Vsesojuznoje Agentstvo Po Avtorskimpravan, for the right to reproduce *Russian Fighting Geese* by photographer Eazhigailov.

Joseph Batty, Beech Publishing House, for permission to use various charts and findings on the source of domesticated poultry and permission to quote material from the Sir Edward Brown book.

Jeffrey H. Black for permission to reproduce photographs of many chicken breeds.

Professor Xiangpin Qui for permission to reproduce his article, *China.*

Dr. A. M. Guhl for permission to reproduce his article, *The Social Order of Hens.*

Dr. W.D. Gross, Virginia Polytechnic Institute, for permission to reproduce his article *Operation for Devoicing Peacocks.*

E. F. Horswell, J. Hazandras, the Sladmore Gallery, London, England, for permission to reproduce photographs of 19th Century and contemporary sculpture relating to poultry.

Lastly, I would like to thank Silvio Mattacchione for his persistence.

INTRODUCTION

The concept for this book originated in February, 1961. I went to visit Mr. Thomas Cleworth on his farm near Milledgeville, Illinois, to meet him and see his Wyandottes. Mr. Cleworth was a very friendly man and I told him of my dream to write a book about the poultry of the world. I will be forever grateful for his encouragement.

In 1972, I began a worldwide search for information and pictures. I gathered material by letter and phone calls. I contacted different nations, their Embassy's in Washington, DC, Ministries of Agriculture, and libraries, following each referral. At times, this was a very frustrating process. However, I did receive exceptional cooperation from many people who responded to my mailing of over 9,000 letters. Representatives of many nations took photographs to be used in *Poultry of the World.*

I launched a worldwide poultry picture contest offering $2,500 in prize awards and received 4,000 photographs. I purchased many pictures from professional poultry photographers in England, Germany and Japan and reviewed 7,000 photographs and illustrations to make the final selection for this book.

Poultry of the World is the ultimate milepost in my life and though it presents many breeds, it is impossible to show every species of domestic fowl, because for each large fowl, there is also a miniature, the Bantam. I am not a poultry breeder, nor an exhibitor, nor a geneticist. My background is the commercial field of the poultry industry. It is simply my wish to share the pictures and information I've gathered through my research.

Above
Claimed to be Biblical chickens in Israel

Below
Iraq Rooster

Left
Eggs, Chickens and Turkeys being sold on the street markets of Ghana

Above Left
Libya Rooster

Above Right
Eggs for sale in Saudi Arabia

Left
Old hens are often preferred to young broilers in Damascus, Syria

Opposite
Selling boiled eggs in Nigeria

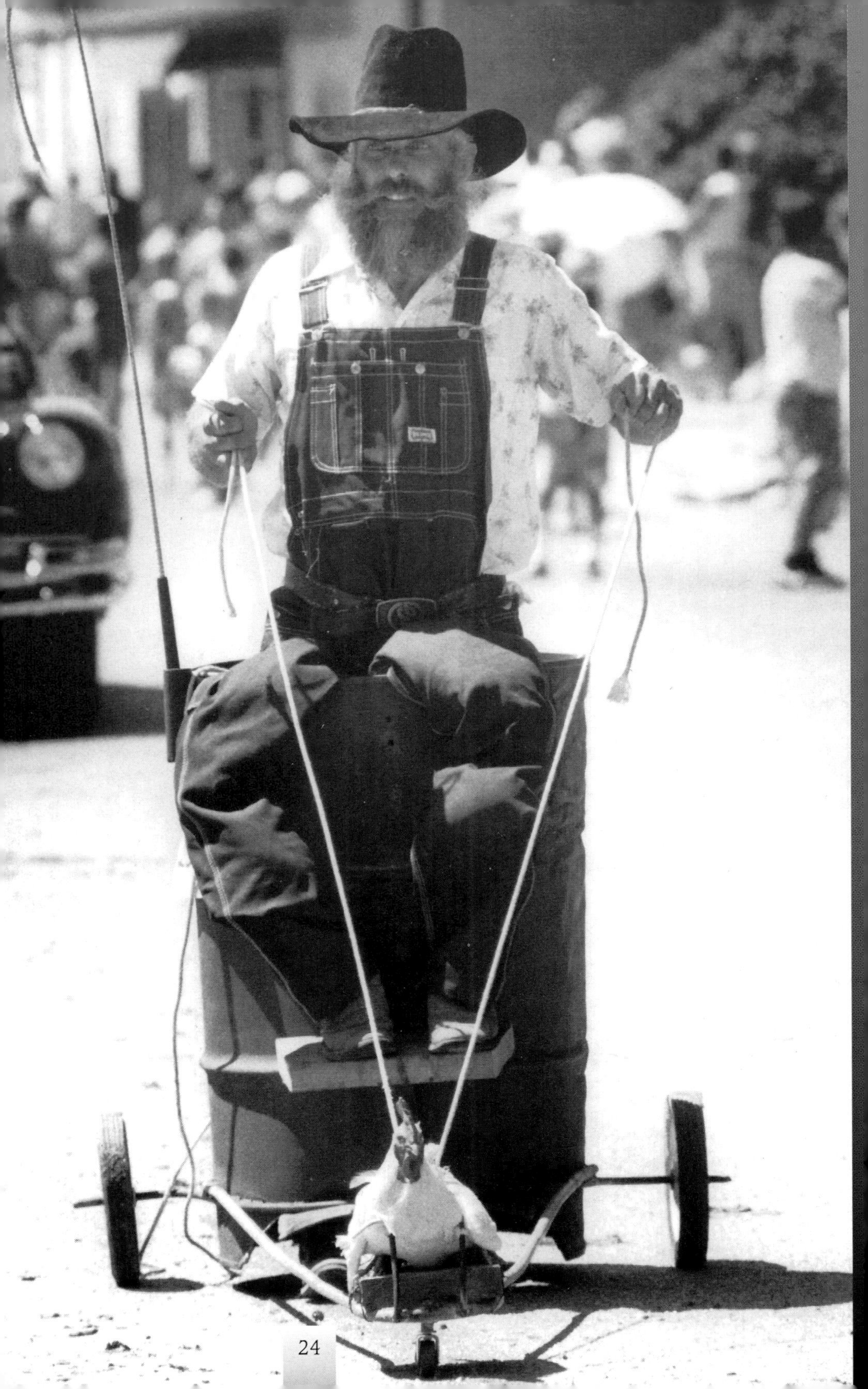

Opposite
Ron Staufer of Nebraska
.S.A. in parade with hard
working rooster

Right
eg Tutty of Australia with
vinning Old English Game

Below Right
hickens being sold in the
public market of Yoemen

Below
Ethiopian Chicken

Peddling his chicken
in street market of
Saudi Arabia

Opposite Page
India, Mobile charcoal chicken conveniently available

Vasalou Rou's campaign to provide a boiled egg once a week for school children

Bangladesh poultry house

Right
Hen mothering puppies
U.S.A.

Below
Mother Sulmater with chicks
Germany

Opposite Page
Hans Schippers of Holland presenting a silver plate as remembraces of his visit to us in 1987. Hans has made several trips to the U.S.A. to bring back new breeds. I call him the Poultry Ambassador of the World.

"Pidan", a special egg treat in Taiwan

Top
Gathering eggs on commercial poultry farm in Syria

Above
Long Island "Big Duck"
20 feet high, 30 feet long.
U.S.A.

Left
Chickens and pigeons being sold in a public market in Ghana

Top
Selling baby chicks on street corner in Damascus, Syria

Below
Egg retailer in Damascus, Syria with hunting falcon for sale

Top
Mr. Moravcova with his large Brown Leghorn hen
Czechoslovakia

Middle
Native chickens of Bangladesh

Bottom
Sliced turkey tails being sold in public market of West Africa

I Love My Chicken

Above
Grandsons Carl and Eric with 8 foot high rooster near Two Harbors, Minnesota, U.S.A.

Above
Helmut Volk from Germany with pet Welsummer

Above Left
Autralian Prize Winners

Below Left
John Impe from Belgium with "Taiwan Shamo"

Above Left
David Pownall from England with Black Cochin Bantam

Above Right
Much patience for this picture, U.S.A.

Below Left
Paul Carter from England, training his Silver Sebright to pose

Below Right
Dr. Wise's daughter from U.S.A. with Belgian Porcelain Booted Bantam

Above Left
Drying bantam for show, France

Above Right
Philippa Black from Germany with Black Rose Comb
Photo: Jim Hare

Below Right
Perfect timing, Germany

Over the past fifteen years, I've collected a vast number of poultry books, many of which are very old editions, now out of print. My oldest book is *The Discours Oeconomique* by Prudent Choyselat, written in 1580. I also have *Aldrovandi on Chickens*, a Latin book written in 1600 and translated into English, and a French book, written in 1750 by M. de Reaumur *The Art of Hatching and Bringing Up Domestic Fowls.*

I received many books as gifts from Poland, Hungary, Russia, Czechoslovakia, East Germany, Holland, Spain, Yugoslavia, Egypt, Thailand, Pakistan and Vietnam. One exceptionally rare volume, *Poultry Farming as Described by Writers of Ancient Rome* was sent to me from Professor Sergio Perdisa, Italy. It was published by the Ministry of Agriculture in Italy to be offered at the seventh World Poultry Congress of 1939, held in Cleveland, Ohio. This book is sprinkled with quotes from the first writers on the subject of poultry, Cato, Varro, Columella and Palladius and is written in both Italian and English. The Stromberg library is unique because most of the countries of the world are represented. I hope you will enjoy, *Poultry of the World.*

Opposite Page
'Poultry Around Ancient Ruin'
by
Auguste Cain
French (1821-1894)
The Sladmore Gallery

'Cockerel and Snail
by
Ferdinand Pautrot
French (1832-1874)
The Sladmore Gallery

Left
'Rooster and Lizard'
Ferdinand Pautrot
French (1832-1874)
The Sladmore Gallery

FROM THE DESK OF SILVIO MATTACCHIONE

Loyl Stromberg's life parallels one of the great industries of our nation, that of the development of poultry and its history, of which this book will now become a part.

Stromberg's book has tremendous value for the poultry breeder, fancier and exhibitor. It is with pride that I present, *Poultry of the World.*

Opposite Page
'Canada Goose'
by
Geoffrey Dashwood
English, Contemporary
The Sladmore Gallery

Below
'Pair of Turkeys'
by
Christophe,
French, Late 19th Century
The Sladmore Gallery

Above
'Poultry Around a Basket
by
Auguste Cain
French (1821-1894)
The Sladmore Gallery

Opposite Page
'Teal'
by
Geoffrey Dashwood
English, Contemporary
The Sladmore Gallery

Chapter One

History of Poultry

Opposite Page
'Cockerel'
by
Geoffrey Dashwood
English, Contemporary
The Sladmore Gallery

Extinct Chickens

When the first *American Standard of Excellence* was published in 1874, some breeds that had existed up to that time, gradually disappeared. Lack of recognition for many breeds meant that they lost favor with poultry fanciers in the United States. But, this does not apply to the breeds which disappeared before that time.

The demise of breeds in Europe raises the question of how many of our breeds came into existence, considering that Professor Ullisse Aldrovandi wrote his book *Aldrovandi on Chickens* in 1600. Yankee Games, English Ravens, Bavarians, Bolton Greys and Guelderlands were all to be exhibited at the first American Poultry Show in Boston and would, within a relatively few years, become extinct.

The following is a partial list of chickens that have become extinct:

Coventry Whites
Polecats
Lincolnshire Buffs
Jersey Blue
Grey Livonian
Paradise Fowl
Blue Dunn Fowl
Domburg
Famenne
Huttigem
Modave
Izegems Fowl
Wherwell
York
White Wonder
Vierlander
Dieman
Suisse Graubunder
Scotch Grey
Cuckoo
Bucks County Fowl
Sherwoods
Hapengouwse
Seloigne
Wyndham
Orolwitzer Buffs
Lomotta
Marsh Daisy
Custards
Dunghill Fowl
Rumpkin
Ostrich Fowi
Guelder
Shackbags
Vreeburg
Flandre Cuckoo
LaPlaigne
Zottigem
Oldenberger
Surrey
Hornet
Hittfelder
Luneburger
Nassuws Lege
Maggi
Gingers
Black Russian
Lark Crested Fowl
Blue Zundert Fowl
Kortrijks
Essex
Mendel
Dimarslandhuhn

In Search of Biblical Birds

By Reverend Ray Trudgian, MA, FRSA
(Reprinted with permission from Poultry Overseas)

Domestic poultry are not mentioned in the Old Testament, I read in one of my attempts to trace the origin of our domestic fowl. No crowing in the Garden of Eden? No cock and hen marching in two by two into Noah's Ark? I poured over Bible dictionaries and Hebrew lexicon. It was true. The only mention was of fatted fowl in the book of Kings which were required by Solomon for his table. As Solomon's kingdom stretched to Egypt and N. Africa it is argued that these were fatted African geese, a favorite table bird as described in our older poultry books.

It is not until the New Testament that I came across two specific references to a hen and a cock. The hen with chicks was seen by Jesus as He stood on the Mount of Olives and wept over Jerusalem. *How many times have I wanted to put my arms round all your people, just as a hen gathers her chicks under wings, but you would not let me.* As a countryman from Galilee Jesus must have warmed to see the hen and chicks on the hilly outskirts of the city and it provided the perfect illustration. Unfortunately the word used for hen gives us no clue as to what sort it was as it is the normal Greek word *ornis* (bird) from which we get ornithology etc. Today the spot is marked by a chapel which has a beautiful mosaic on the altar, of a dark game type hen with outstretched wings, a halo surrounding the head and chicks nestling underneath.

Opposite Page
Portrayal of rooster that crowed three times when Peter denied Christ in famous Church in Barcelona, Spain

The cock is referred to as *alektor* and is said to have crowed when Peter denied his Lord on the night of his trial as Jesus said he would. The Romans ruled Palestine then and they had a guard change at the third watch in the middle of the night at 3:00 a.m. which was sounded by a bugle called *The Cockcrow*, and

GAL
LVS

scholars have argued as to whether Peter heard an actual cock or the Roman bugler but they all seemed agreed that cocks start crowing at 3:00 a.m. in the Middle East! This spot now has a splendid church which the Crusaders called St. Peter in Gallicantu (at the cock crowing). Poultry breeders probably know that *Gallus* is the technical term for our birds and cantu is linked with chant-singing.

My wife Angela and I had a chance to visit Israel for eight days in January '86 and we set out in search of the native birds as we went on our pilgrimage. We seemed to be on to a non-starter as farming in Israel is highly organized into the *Kibbutz* or community farm system with European-type turkeys, poultry and geese raised by semi-intensive straw yard husbandry. Turkey and chicken graced our table daily - it was just like being back in Norfolk!

Visiting Bethlehem which is largely an Arab town now, I went to the back of a souvenir shop and heard a cock crow. With great excitement my wife and I crept round the back to be met by a pack of yelping guard dogs but we were able to see that they were protecting a splendid flock of what could only be golden Sicilian Buttercups - the magnificent buttercup comb was unmistakable and the upstanding profile was as I remembered it from the photo in E.T. Brown's Poultry textbook. Easom Smith in his *Modern Poultry Development* would seem to confirm our amateur findings as he traces the origin of this (wrongly named) breed to the Arabs of North Africa who would have been moving freely in the Middle-East in New Testament times and describes the cup-shaped comb as a *veritable King David's Crown.* Perhaps it was fitting that we found our Biblical bird in Bethlehem, Royal David's City, for according to Christian legend it was in that city on that first Christmas Eve that the cock was the first to announce the Birth of Christ. Instead of crowing *Cock-a-doodle-do* he sang CHRISTUS NATUS EST (Christ is born). The words fit the tune - try it sometime!

Why is a Chicken Called a Chicken?

By Reverend Ray Trudgian, MA, FRSA

Perhaps we can begin to answer it by asking another question: *What is a chicken?* Nowadays we use the word to cover all kinds of fowl but strictly speaking it was applied only to the baby chick in its fluffy state before it gets feathers and can be recognized as a pullet or a cockerel.

Originally the word *chick* was the singular and *chicken* without the *s* was the plural as in *ox* and *oxen.* In my native West Country the word is still used in the way in such phrases as *he keeps pigs and chicken in his backyard.*

Although these strict rules are no longer widely observed they help us to go back to the root of the word. It is certain that the Ancient Britons were trying to imitate the *cheep cheep* sound of the baby chick when they called them in Old English *cicen* in the singular and *ciceu* in the plural. In Medieval English about 1300 as the word had become *Chikenes.*

The sound of these words is similar to the Dutch *keiken* and the German *kucklein* which leads some scholars to trace all the words back to an Old Teutonic root *kuidino* which also gave rise to the word *kuk* or *cock* referring to the special cluck of the cock bird. As a missionary in Central Africa it was interesting to note that the various tribal languages called domestic fowls *nkuku* (singular) and *cikuku* (plural) which again relates to the clucking noise and the universal language of our domestic fowl brings together ancient European and African languages.

An interesting chronological study of the evolution of the word *chicken* can be seen in the various translations of St. Matthew's Gospel, Chapter 23 verse 37 where Jesus speaks of gathering the people of Jerusalem to him as *a hen gathers her chickens under her wings* (King James Authorized Version 1611).

"Two Cocks Carrying A Fox On A Stick A 12th Century mosaic on the floor of St. Mark's Church in Venice showing that cock and fox stories were very popular in Europe some 200 years before Chaucer wrote his tale of Chanticleer and the Fox. Chaucer paid a number of visits to North Italy in the service of King Edward III. Illustration donated and collected by Rev. Ray Trudgian who was impressed with the many detailed paintings of poultry in art galleries throughout the country. This mosaic is 500 years older than the Aldrovandi paintings.

The earliest Gospels were written in Old English by some monks off the coast of Northumbria called *The Lindisfarne Gospels* in 950 AD which refers to a *henne* and *hire cicceno*. Another translation in 975 AD shows that the word has already developed into *ciken* and by the year 1000 AD the translation refers to *henn gathering cicenu under hyre.*

Chaucer, who gave us the wonderful picture of domestic poultry keeping in the story of Chanticleer and the Fox in the 1380s uses the term *chikenes* rather broadly and by the time we get to Shakespeare in the 1590s and the writing of the Authorized Version of the Bible at about the same time we already have our modern word *chickens.*

Above
1700 year old cave for chickens in Syria with their Baldy chickens. Most likely the only cave chicken shelter in the world.

Below
Copy of first painting from 2620 B.C.

Opposite Page
Rev. Trudigian with pet Maran

It is interesting to note that some modern versions of the Bible translate the word as *chicks* but the American Revised Standard Version, 1951, goes back to the original Greek and translates it as *brood* from the Greek word *nossia* which gives us the word *nest.* The Greek for *hen* is *ornis* from which we get our word *ornithology.*

It may be of interest to note that related words like *fowl* come from the Old English *fugel* linked with the word meaning *to fly* and the term *poultry* has developed from the languages around the Mediterranean, Latin *pullus,* Italian and Spanish *pollo* and the Old French *pouletrie* and Modern French *poule.*

Could it be that our modern wide use of the term chicken is influenced by our eating habits? KENTUCKY FRIED CHICKEN sounds much better than KENTUCKY FRIED FOWL!

As a traditional West country poultryman for me a *chick* will always be the fluffy thing that bursts excitingly out of the shell each spring and *chicken* means lots of them--the more the merrier!

Chronology of Poultry History

3000 B.C.	Four species of wild chickens were known in south and southeast Asia, along with the first domesticated chickens.
2000 B.C.	Domesticated chickens were found in the cities of Mohenjo, Daro and Harrapa of the Indus culture in Pakistan. Cockfighting was described.
1500 B.C.	In China and Egypt, domestic poultry from India was found, origin unknown.
1200 B.C.	Code of Manu prohibits the eating of village poultry and domesticated fowl. They were allowed to eat wild fowl only.
1000 B.C.	Domesticated poultry appeared in all cultures of the Middle East except Israel. Artificial hatching was known in China and Egypt.
800 B.C.	Domestic poultry appeared in countries around the Mediterranean and regions of Europe.
600 B.C.	Domestic poultry appeared in Babel and Assyria after the Banishment.
520 B.C.	First known Greek pictures of chickens, especially of fighting cocks, depicted on vases, altars and tombs.
480 B.C.	Themistocles introduced official cockfighting in Athens.
390 B.C.	Ancient Rome was saved by honking geese which alerted them to the impending invasion of Gaels.
200 B.C.	Romans mentioned poultry and cockfighting. Most famous rooster in the world crowed when Peter denied Christ three times.
A.D. 1-100	Varro, Plinius and Columella wrote about poultry keeping, giving the names of eight breeds.
486	*The Lex Salica* mentioned the punishment for stealing poultry.
768-814	Charlemagne decided the number of poultry that farmers must keep and installed Plumgraven, Feather Count of Earls, to record whom kept poultry in his name.

Pharaoh Quail
Bas Relief from Tomb of Mereruka
Pharaoh, Mereruka's peasants netting quail in wheat fields about 3,000 years before Christ. From Mastaba of Mereruka, University of Chicago Oriental Institute Publications

Ancient picture of dressing out Egyptian Geese

Painting of a cock on a potsherd, from the tomb of Tutankhamun, Egypt. 1350 B.C.. This presents the possibility that the early stock for Leghorns credited to Italy may have come from Egypt.

1096-1270	During the Crusades, poultry breeds were exchanged in Europe.
1475	Peter Jasperz, Holland, had white poultry with black crests.
1500	Europe began exporting poultry to America and South Africa.
1512	Painter Durer created a painting of a crested cock.
1580	First chicken book was published in French, *The Discours Deconomique* by Prudent Choyselat, translated into English in 1951.
1600	*Aldrovandi on Chickens*, by Ulisse Aldrovandi, was published in Latin. This was the first significant poultry book and was translated into English in 1965.
1626-1679	Jan Steen, Holland, painted animals and poultry as did Albert Cuyp, Monckhort and Melchior 'd Hondecoeter.
1749	M. de Reaumur published his book, *The Art of Hatching and Bringing up Domestic Fowls* and predicted profits in poultry.
1775	The Netherlands and Belgium published the first professional books on poultry.
1788	George Washington imported the first geese into the United States from China.
1828	Leghorns were imported into the United States.
1835	Java chickens were imported into the United States.
1840	First official poultry census was taken in the United States.
1841	Dorkings were imported into the United States by L.F. Allen, England.
1843	First patent was granted for an American incubator to Napoleon Guerin, New York. Cochins were imported into England from China. First American poultry book by C.N. Bement, *American Poulterer's Companion*, was published.
1845	First known poultry show was held in Regents Park of London, England and was sponsored by the London Zoological Society.

1846	Brahmas were imported to the United States. Queen Victoria of England exhibited a pair of Cochins at the Dublin Cattle Show.
1848	Popularity of the Cochins was an influencing factor of the first organized poultry show at Birmingham Hall, Birmingham, England.
1849	Cockfighting in England was outlawed by Queen Victoria. First American Poultry Show was held in Boston, Massachusetts.
1850	First poultry breeders association was formed in Boston, Massachusetts. This became the American Poultry Society in 1859. Large scale duck raising was begun in Long Island, New York.
1856	Barred Plymouth Rocks breeding development was begun.
1865	First English Poultry Breeding Standards book was published.
1867	Belgium prohibited cockfighting but it continued in secret.
1868	Crystal Palace Show was held and has become a classic show in England.
1869	First trapnest patent was granted to D.P. Liech.
1870	*American Poultry Monthly* was published by the New York Poultry Society. Jersey Black Giants were developed by the Black brothers, New Jersey. Toe punch was introduced as an aid to selective breeding.
1871	One hundred people attended the first special banquet for poultry breeders organized by the New England Club of Worcester, Massachusetts.
1872	First crossing for improvement of egg production was done by T. Timms, England.
1873	American Poultry Association was organized at Buffalo, New York. First Pekin ducks were imported into the United States from China.

1874	*American Standard of Excellence* was published. Barred Plymouth Rocks and Dominques were admitted. First White Plymouth Rocks were introduced.
1876	Felch, Nadick, Massachusetts, published his poultry book, *Poultry Culture.*
1878	Egg drying process was developed in 1865 and became a commercial operation.
1880	J.C. Campbell built his first Eureka incubator.
1881	First mechanical refrigeration for butchered poultry was built in Boston, Massachusetts.
1883	Java chickens were imported into the United States. C.E. Hearson became the first British inventor of a self-regulating incubator.
1884	Monarch incubator was manufactured by James Rankin.
1885	White Wyandottes were introduced by Reverend G.M. Briggs, Wyandotte, New York.
1886	First advertisement of one dozen paper egg boxes appeared.
1888	Prairie State incubator was advertised. It was successful, and stimulated artificial hatching in the early 1890's. First brooder stove was advertised to be used with a homemade hover.
1889	Gas lighting was used to stimulate egg production.
1890	First Madison Square Garden Show was held in New York city. First trapnest was patented.
1892	Long distance shipment of baby chicks went from Stockton, New Jersey, to Chicago, Illinois. First poultry course was taught by James Rice at Cornell University.
1893	Silver Laced Wyandottes were admitted to *American Standard of Excellence.*
1894	First Royal Ornithophillia Show was held in Holland.

1895 First issue of *Poultry Tribune* magazine was published in Freeport, Illinois.

1896 A 20,000 egg capacity was built by Charles Cyphers Company. Sir Edward Brown helped to organize the Utility Poultry Society of England, which became the World Poultry Science Association in 1912.

1897 The British Utility Club sponsored the first egg laying competition in England.

1898 Rhode Island State College offered the first short course on poultry.

1899 Hand-operated, full or hard scald, became popular for poultry dressing. Pullorum disease germ was isolated.

1900 First shipment of baby chicks was recorded in Canada. Thirty-six dozen egg cases were replaced by thirty dozen cases.

1901 Dry mash feeding was developed.

1902 Dutch organization, VPN, was organized to promote poultry products. 200,000 incubators were exported from the United States.

1903 Gasoline brooder was developed.

1904 Kansas State College held the first official egg laying contest in the US. Rhode Island Reds and Buckeyes were admitted to *American Standard of Perfection*. National Serum Institute, Rotterdam, was founded to fight poultry diseases.

1905 Alfalfa was used in feeds. First battery brooder was designed by Charles Cyphers. Belt feeding system was patented by G.C. Scott.

1906 Spelderholt Research Institute of Holland was established in Beekbergen. Cyphers Incubator Company developed a double-deck incubator.

1907 Roy Slocum became the first poultry researcher in the US Department of Agriculture.

1908	Electric candler was developed and a coal burning brooder developed.
1910	Average egg consumption per person exceeded 300 eggs in the United States. Fowl vaccination was perfected.
1912	World Poultry Science Association was organized in London by Sir Edward Brown. First 300 a year egg-layer, a White Leghorn, was developed at Oregon State College. General Poultry Association, Rotterdam, Holland, published a magazine for poultryman.
1913	In Barnveld, Holland, the first egg exchange market was organized. First oil brooder was advertised by Petaluma Brooder Company.
1914	*Poultry Press*, a newspaper, was established by the DeLancey family, York, Pennsylvania. Experiments were begun with an electric brooder. USDA issued a bulletin with instructions on how to ship hatching eggs by parcel post.
1916	International Baby Chick Association was organized.
1917	Baby chicks were admitted to US via the Postal service. South Dakota passed the first *good egg* law. Manufacturers of baby chick boxes agreed to standardize sizes. Pullorum eradication program was originated by Professor O.W. Richardson, New Hampshire, whom became active in the development of New Hampshires.
1918	Vaccination for chicken pox began at Vineland, New Jersey Egg Laying Contest. The use of batteries was recorded for laying hens and growing pullets. Smith incubator was patented with a 47,000 egg capacity and the first incubator to use forced draft principal was manufactured.
1920	First advertisement of the Buckeye mam moth incubator appeared. A scientific study at the Massachusetts State

	University was done to establish how often a hen moved her eggs. Observation showed they were turned as often as five times an hour.
1921	First World Poultry Congress was held in the Hague, Holland, and NPF was founded in Holland. It was a federation of clubs for professional poultry breeders.
1922	Rhode Island Whites and Jersey Black Giants were admitted to *American Standard of Perfection.*
1923	Electrically heated, large capacity incubators were invented by Ira Petersime. Dr.W.T. Johnson reported his findings on the disease, Coccidiosis.
1924	Wisconsin introduced *all mash* chick ration. Cod liver oil, an early source of vitamins, was used in commercial starting mashes.
1925	Infectious laryngotracheitis was reported. Semi-scald method of poultry picking was introduced.
1926	Fowl pox vaccine was available. Started chicks, three to six weeks old, were offered to the public. A monument to the Rhode Island Red was erected in Adamsville, Rhode Island. Harry Kettle of Dallas, Texas, flew a chicken to Madison Square Garden Poultry Show for exhibition.
1927	Chick sexing was demonstrated at the World Poultry Congress in Ottawa, Canada. Battery brooders received public attention. Dent Holcomb, Nebraska, advertised baby chicks on the radio.
1928	Sex-linked crosses of chicks were advertised.
1929	Bundy Incubator Company introduced the separate hatcher method of incubation. Blue Comb disease was reported by Dr. F.R. Beaudette, New Jersey. Dutch Egg Contrition Station was founded.

1930	Pullorum testing showed rapid progress with commercial hatcheries. Baby chicks were shipped by railway express. Holland produced 24 million chickens and Belgium, 12 million. Barnvelder chicken declined in popularity.
1931	Laying cages were given publicity. Rubber fingers for mechanical chicken picking was invented by George R. Hunt. First demonstration of chick sexing in US was done at the International Baby Chick Convention in Grand Rapids, Michigan. Laryngotracheitis vaccine was developed by Dr. F. R. Beaudette, New Jersey. An over-production of eggs in Holland resulted in government controls that stayed in effect until 1953. Lamonas were admitted to *American Standard of Perfection.*
1934	Eleven thousand, four hundred new chick hatcheries were listed in the U. S. which was an all time high.
1935	First chick sexing schools were organized offering sexed chicks to the public. Artificial insemination of poultry was introduced. First of the sulfa drugs was introduced. New Hampshires were admitted to *American Standard of Perfection.*
1936	In Barnveld, Holland, a second egg exchange was built with the claim it was the greatest free egg market in the world. First sexing of turkey poults was done.
1937	Inbreeding of poultry was begun and led to hybridization.
1938	Automated poultry processing was developed by the Barker-Gordon Johnson company.
1940	Era of mechanical poultry dressing was begun when small pickers were developed. Beltsville White turkeys were developed by US Department of Agriculture. HyLine Poultry Farm marketed their first Hyline inbred hybrid chicks.

1941	Newcastle disease was reported by Dr. J.R. Beach, California.
1941	Electric debeaker was developed by Lyon Electric Company, California
1943	A Rhode Island Red, bred by E.B. Parameter of Massachusetts, produced 351 eggs in 357 days at Western New York Egg Laying Contest.
1946	*Chicken of Tomorrow* contest was launched by Great Atlantic and Pacific Tea Company, beginning a tremendous improvement in the broiler industry.
1948	Live virus vaccine was developed for the control of Newcastle disease. *The Poultry Tribune* circulation totaled 500,000 sub scribers.
1949	Mechanical egg gathering device was developed in New Hampshire and a random sample egg laying contest was started.
1950	Automatic egg breaking machine was developed by Seymour Packing Company. Trapnesting of geese was done at University of Minnesota.
1952	American Poultry Historical Society was organized. Coccidiosis vaccine was announced by Alabama Experiment Station. Delawares were admitted to *American Standard of Perfection.*
1953	This was the end of egg quota intervention by the Holland government.
1954	As a result of over- production, egg prices dropped dramatically in Holland.
1955	A new era of egg production breeding using Reciprocal Recurrent Selection was begun. Clear and colored plastic egg cartons were developed by Cornell University.
1956	Colonel Sanders introduced Kentucky Fried Chicken and increased the popularity of chicken within the fast food industry.
1957	Dr. Briles, Texas, announced the

	significance of blood typing , introducing a new poultry frontier.
1959	Stahlman Farms, New Mexico, marketed over 20,000 geese.
1961	Cage laying stems expanded in the Midwest and Northeast.
1964	Pullet raising became a specialized business.
1965	Egg consumption dropped and prices were depressed due to a cholesterol scare.
1966	*NOP*, Netherlands Organization for Poultry, published their weekly magazine, *De Plumveehouder* (Poultry Keeper).
1970	Holland had its last outbreak of Newcastle disease resulting in a very strict poultry health program.
1971	An average of 322 eggs per person was consumed in the United States.
1971-73	Newcastle outbreak in California. Cost to the Federal Government over $50 million.
1972	Marek's vaccine introduced in 1972. Mortality rates dropped by 13 percent by 1976 and even more in subsequent years.
1972	Watt Publishing Company estimated that 75 percent of the layers and 21 percent of the pullets in the U.S, were housed in cages.
1972	A Watt Publishing Company survey revealed that some 38 percent of the egg output in the U.S. was produced under some form of contract. (It wasn't many years before this percentage dropped as egg production complexes were established).
1974	Book: "American Poultry History" published by the American Poultry Historical Society.
1975	This year ended the longest running egg cost study conducted in California starting in 1925.
1975	The American Egg board (AEB) was voted into existence, effectively replacing the Poultry and Egg National Board. AEB has

the power to enforce mandatory assess ment from egg producers.

1980 United Egg Produers writes Welfare Guidelines for egg producers, the first animal industry to do so.

1983-84 Avian infuluenza in Pennsylbania region. Cost to U.S. Government over $60 million. American Poultry Association publishes first Color Standard of Perfection.

1984 The Egg Nutrition Center (ENC) was created by the egg industry to provide scientifically accurate information on egg nutrition and accompanying health issues.

1985 West Coast United Egg Producers was established as the fifth Regional member of United Egg Producers.

1986 Salmonella enteritidis outbreak in Pennsylvania. Later spread to many other states, primarily in the Northeast. Poultry Press is purchased from Bob Delancey by Bill Wulff.

1988 The California Egg Marketing Association was established (under the Capper-Volstead Act of 1922) as the first known egg marketing cooperative established for the primary purpose of setting producer paying prices for eggs.

1988 Museum of the British poultry industry opened at Stoneleigh, Warwickshire, England.

1989 A book was published entitled 'Western Poutlry History' under the sponsorship of the Pacific Egg & Poultry Association. Dutch Poultry Museum opens in Barnveld, Holland. U.S. Postal Department makes available Express Mail for shipping adult poultry.

1989 Egg product production reached 19 percent of total egg sales.

1991 A survey determined that there were

approximately 156 in-line egg production complexes in the U.S, with 46 percent of the nation's laying flock. (It was forecast that this number would increase to 220 complexes with two-thirds of the national flock by the year 2000, with 60 of these housing one million or more hens).

1991 A survey revealed that some 63 percent of pullets in the U.S. were being reared in controlled environment houses compared to 71 percent of the layers.

1991 A University of California survey showed that 97.8 percent of the layers and 70.6 percent of the replacement pullets were housed in cages.

1992 "A Review of the Cholesterol Controversy is the United States" was prepared by A. William Jasper, summarizing the relevent events in the cholesterol wars of the previous 40+ years.

1993 University of California studies of 3,586 commerrcial U.S. white Leghorn laying flocks hatched between 1985 and 1993 show remarkable progress in various performance traits.

1993 Roland Hartman, world-renowned poultry editor, and author of *Keeping Chickens in Cages*, a book that established a base for worldwide cage egg production, retired at the age of 87 for the third time, this time from Inland Empire Poultrymen, a trade association headquarters in Redlands, California.

1994 In 1994, the nation's hatcheries produced 379 million straight-run chicks, an average of about 15.5 million pullet chicks per month. By this time, there were fewer than a dozen major hatcheries of significant size and another dozen smaller hatcheries serving the entire nation's egg industry.

November 30th- Ribbon cutting dedication of 'phase one' of the National Poultry Museum.

1995 Egg City in Moorpark, California, once considered to house the largest egg flock in the world *(ultimately reaching more than three million layers)* was set to close down permanentely.

1995 The Leghorn business in the U.S. is now divided *(unequally)* among only three major breeders, and all of the more popular strains are owned by foreign countries.

1995 Egg product production reached almost 30 percent of total egg sales.

1995 United Egg producers provided financial support to enhance efforts of
(1) the Animal Industry Foundation regarding the animal Welfare-Animal Rights controversy;
(2) AG In The Classroom regarding agricul tural education for teachers in schools throughout the country, and
(3) Project Food, Land, and People for the purpose of developing student curricula for schools.

1995 United Egg Producers developed the '5-Star' Total Quality Assurance Food Safety Program.

1995 Alliance established between the United Egg Association and the United Egg Producers making it possible to provide leadership for all segments of the egg industry so that all facets of the industry can speak with a unified voice on legisla tive, regulatory, and other important matters.

1996 Poultry of the World by Loyl Stromberg is published.

We credit Dr. Bill Jasper and Professor Donald Bell of California for the Chronology 1971 through 1996.

DORKING COCK

By permission of the "Field"
VERY YOUNG LIGHT BRAHMA COCKEREL, 1853

THE DORKING FOWL.*

WHITE DORKING COCK.

Whom Should Take Credit for Origin?

Regarding origin, the first breed to come to mind is the Cochins, which have long been recognized as having originated in China, which they did. They were vastly different than the Cochins of today. However, the English should be given the first credit for their development. Americans joined in their improvement to bring out the fluffy plumage and massive type. The English and American's developed the present Cochins.

Italy is the source of the first Leghorns. It is apparent that they did not undergo the change that Cochins did. The first Leghorns that were imported were thought of as Brown or Black. Most of the colors, except for the Buff Leghorns from England, were developed in the United States.

Mexico was the original homeland of the first wild turkey. The Spanish brought them to Europe. Most of the colored turkeys were developed in the United States. A few breeds were developed in Europe.

Though there is still the question of which is truly correct, I present the breeds as other poultry books have done...from the country of origin.

HER MAJESTY'S COCHINS, IMPORTED IN 1843.

Chapter Two

Myths of Domestic Poultry

Aldrovandi on Chickens

Aldrovandi on Chickens, was written in 1600 by Professor Ulisse Aldrovandi and was the first significant work on poultry. It was published in Latin and translated into English in 1963, by Dr. L.R. Lind at the University of Oklahoma. Though this was not the first poultry book published, I refer to it as the first significant one.

In my research library I have a copy of the first book published on poultry, *The Discours Oeconimique* written by Prudent Choyselate in 1580. However, this book does not compare with the valuable information that Aldrovandi provides. Another book entitled, *The Art of Hatching and Bringing Up Domestic Fowls,* written in French by M. de Reaumur in 1750, though exceptional for its time, does not compare well with Aldrovandi's book of 1600.

Professor Ulisse Aldrovandi was one of five great naturalists of the 16th century and established the Museum of Biology. In 1549, Aldrovandi and seven other intellectuals were accused of heresy. They were sent to Rome for questioning but were released by Pope Julius III. Not satisfied with his freedom, Aldrovandi stayed in Rome until he was completely exonerated.

At the urging of his family, he married at the age of 41. His work was his priority during his entire life. In the year 1600, after 46 years of teaching, Aldrovandi applied for retirement and was granted a pension equivalent to that of his salary. He passed away three years later at the age of 83.

Gallina Patavina
(Istituto de Zoocolture - Università di Bologna

Gallina e Gallo Grifutis pedibus

Gallo con cresta a cornetti

Gallus Spado

Gallus Persicus sine cauda

Gallus Turcicus

Gallina Patavina ex albo virescens
Melanochloros capite cirrato

Gallo e gallina Padovana
(Istituto de Zoocolture - Università de Bologna

Indian Hen

Wool-bearing hen
with rooster's crest of the botanists

The swamp rooster

The Male Paduan

Aldrovandi was an outstanding zoologist and produced several noteworthy manuscripts. The value of his work was not immediately recognized. Many manuscripts were published after his death. He frequently quoted from other writers who had preceded him in order to expand the value of his own work. These included Cato, Columella, Varro, Marcellus, Palladius, Socrates, Hypocrites and Plato.

His book is amazing for that era. Here are a few highlights from *Aldrovandi on Chickens.*

Ancients measured time by the crowing of the cock prior to sundials.

Aristotle and Aldrovandi agreed that the chick embryo developed in the white of the egg.

Chicken fat and goose fat (grease) were used for the complexion, to beautify the face.

Cock fights were used to encourage soldiers.

Culling by the ancients was done by girls.

The disease, corzyza, was known and called by the name of pip.

Eggs were incubated in Livia Augusta's bosom.

Four day old eggs are best to eat.

Eggs to hatch should be from hens, not from pullets.

Fox is the real enemy of the chicken.

Hens hate thunder. It destroys the eggs and chick embryos.

Hens foretell rain.

Hypocrites said chicks develop in the yolk of the egg.

Improve lust by eating rooster testicles.

A homemade incubator was heated with a lantern.

Law was passed in Rome to regulate fattening and serving poultry.

Raw eggs are used to treat a sore throat.

Raw eggs are used to treat burns.

Rooster treads hens 50 times a day.

Old roosters are more flavorful when cooked in wine.

Rooster with feathered feet

The Turkish hen with blue feet beside an ustilago plant

Capon with a stalk of alopecuroides

The Persian hen

The white-crested chicken

Social behavior of roosters is called order of the peck.

Shell hardens after eggs are laid.

Wife of Hercules raised many hens and roosters.

There are wool bearing chickens, some with hair (Silkie).

Chicken sculpture in the Hall of Animals in the Vatican Museum

There are only three original sets of Aldrovandi books in the United States. They are in Harvard University Library, National Agricultural Library in Beltsville, Maryland and the Special Collections Department of the University of Kansas.

Aldrovandi on Chickens is currently out of print. I have been encouraging the publishers to reprint it because I consider this book to be a classic that every poultry fancier should own. The English translation was printed in 1963 by University of Oklahoma Press, 1005 Asp Avenue, Norman, Oklahoma, 73019.

Aldrovandi paintings and a statue of a chicken, now in the Vatican Museum, show how chickens appeared in the year 1660.

Domestication of Chickens

Webster's dictionary defines *myth* as a legendary story. I found that this applies to my attempt to reach the earliest period of the domestication of poultry. The most often quoted writings report that the first domestication took place about 2000 B.C. in the Indus Valley of West Pakistan. However, I've also seen this reported around 1400 B.C. Even Charles Darwin agrees that domestication involved the Jungle Fowl. A number of references contend that the first domestication was tied in with cockfighting. Encyclopedia Britannia states the first domestication was reported 5000 years ago and was derived from the Red Jungle Fowl.

Cockfighting history is also confusing. Dr. Clifton Bryant, Virginia Polytechnic Institute and University, suggested that cockfighting may be 6,000 to 7,000 years old. Sir Edward Brown, author of a set of books,

Poultry Breeding and Production of 1929, states the first cockfighting was started in the reign of Croesus, King of Lydia, in 3436. There is more contradiction since the Code of Manu reported cockfighting in 1000 B.C.

Hans Schippers, Holland, reports that there are old Indian manuscripts indicating domesticated chickens in India in 3000 B.C. His research shows an ancient Chinese Encyclopedia in a London Museum reports chickens came to China about 1400 B.C. from India.

Dr. Edmund Hoffman of Canada wrote an article on the early domestication of barnyard fowl. His research indicated the Muscovy duck was domesticated in Peru as early as 3500 B.C. The earliest known painting of six Egyptian geese of 2620 B.C. raises the question of which fowl was the earliest to be domesticated. McGraw Hill's book, *A Dictionary of Birds*, states that the date of the Muscovy duck domestication is unknown, though it was probably in India as early as 3200 B.C. *Evolution of Domesticated Animals,* by Masson and Logman reports evidence of the first domestication in the Harappan culture of the Indus Valley dating from 2500 B.C. to 2100 B.C. It quotes the findings of Zeuner, placing the first domestication at 2000 B.C. Zeuner also speculated that chickens were domesticated first for sport (cockfighting) and secondly, for food. In the *Poultry Science Magazine* published in 1959, an article, *History of the Domestic Chicken from Antiquity to the 19th Century,* reported the first domestication to have been in India around 3200 B.C.

A Russian poultry book from 1962, *Poultry Science, Practice, Biology, Breeds and Breeding*, states that Jungle Fowl were found in archaeological excavations which would indicate domestication in 3250 B.C near the town of Mohenjo Daro, India. They don't know for sure if the Jungle Fowl found was in the wild state or an example of true domestication.

Dr. Roy Crawford, a geneticist at University of Saskatchewan, Canada, in an article entitled, *Conflicting Stories*, refers to the first domestication in the Indus Valley of Pakistan about 4,000 years ago, and thought China might be the oldest culture, developing chickens

Chicken of Pakistan

Chicken of Peru

4,000 years before Pakistan. Domesticated dogs, pigs, and chickens may have spread from China to Europe via what is now Russia.

Bill Plant, a researcher in Australia, thinks the first domestication of chickens occurred about 3500 BC. in Southeast Asia. He reports bone findings in Thailand in 3500 BC. Although there are many reports of chicken bone findings, this is not conclusive. They may have been those wild Jungle Fowl before they were first domesticated.

Most references on the distribution of chickens state the starting point as India. They traveled first to Egypt by sea and overland to Persia, Mesopotamia ,Greece and later, to Italy.

An article by Barbara West and Ben-Xiong Zhou, published in a 1988 issue of *Journal of Archaeological Science* entitled, *Did Chickens Go North? New Evidence for Domestication* , supplies well-founded information. It raises the question on when and where the first domestication of chickens actually took place. The writers present the possibility that it may have been in the sixth millennium in Southeast Asia, then moved north to become established in China by 6000 B.C.

Chickens were introduced to Japan via Korea during the Yayoi Period. There are references to bone findings in many nations; Russia, Pakistan, India, Iran, Turkey, Syria, Israel, Jordan, Greece, Italy, Spain, Switzerland, Spain, Romania, Hungary, Czechoslovakia, Poland, Germany, France, Netherlands, Sweden and Great Britain.

In 1994 more evidence emerged as archeologist Robert Bradwood, University of Chicago, indicated that his research points to the first domestication by Vietnamese farmers over 10,000 years ago. Further reports during 1994 reflected both the similarities and differences in mitochondrial DNA of Jungle Fowls and various breeds of chickens. These reports point to Gallus gallus (Red Jungle Fowl) of the Thailand region as the origin of domestication.

In conclusion, one can see that there are many conflicting records and it is difficult to decide which hypothesis is correct! However, there is agreement that the general area had to be Asia and that the Red Jungle Fowl was the principle stock of origin.

An early picture of a rooster in the 'Book of Animals', produced in Bagdad, Iraq by Al Jahiz in the time period of 800 to 840 A.D. Book discovered in Milan, Italy library in 1939.

Early Writing on Poultry

In 350 B.C., Aristotle stated that long pointed eggs were female and rounded eggs were males and to this day there are those who share in this belief. Varro, from the time of 116 to 27 B.C., was the second to write about poultry. He recommended the following breeds for fighting; Tahagran, Median and Chalcidian which proves an early interest in cockfighting.

Aristotle, Pliny, Varro, and Columella mentioned several breeds famous for their cockfighting ability; Hadrianic Tanagran, Median, Rhodian and Chalcidian. In a book written by Martin Stephenson, England, reference is made to several breeds of miniatures, the Bantams.

Columella, in A.D. 60, came forth with the idea of chickens with five toes. Perhaps they were the start of the Dorkings family. I have found it reported that the Romans brought a square-bodied chicken to England, which may have been early Dorkings.

Darwin writes in his book, *The Variation of Animals and Plants Under Domestication*, that at the commencement of the Christian era, the Romans kept six or seven breeds. He states that in the 15th century, several breeds were known in Europe. China had named seven of these. The Philippines had nine sub-species of game birds or fighting chickens. Darwin also wrote about the black-skin chickens in South America, which would be Silkies.

It is accepted that the Jungle Fowl was the starting place for poultry. However, there is no knowledge that can precisely pinpoint which breed came after this fowl. It might have been some of the fighting chickens. There is speculation that this might have been what Aldrovandi indicated in his book as *Turkish, with blue feet*, which could be the forerunner to Sicilian Buttercups. Speculation exists that after the Jungle Fowl came the Leghorns.

A Russian poultry book of 1962, makes reference to a Leghorn-type chicken known in Egypt in 1450 B.C.. It also has an illustration of a cock bird within one of the Pyramids. Aristotle refers to a similar chicken in the same time period.

I conclude that the Paduan Fowl, as described by Aldrovandi, is the forerunner of our Crested Polish and Houdans. Researchers would like to know when, where, and how various breeds came into existence. All I can say is that ancient people interested in chickens, and without any formal knowledge of genetics, raised an admirable group of breeds that have proven to be a challenge to dedicated poultry breeders.

Economic Value of Poultry in Medieval Times

India and Pakistan are recognized as the areas where the first domestication of chickens took place. In early writings, there is no reference regarding the value of chickens. Cockfighting played an important part in the first domestication, because it was developed into a sport enjoyed by nobility as well as peasants of that era. I can only speculate on the value that may have

Chinese rice-paper painting of men selling poultry. A live cock being weighed and a duck are shown also. This scene illustrates the importance of poultry in the economy of densely populated China; an importance which the fowl acquired in Europe at a comparatively late period.

been placed on an exceptionally good fighting cock. It is likely they commanded a very high price. The exchange might well have been by barter, with jewels, spices, clothing or coin as payment.

The first fighting chickens reached Egypt via trade ships. Nations such as Greece and Persia were the first and later, Italy. Other European countries received their first chickens via over land trade caravans. Again, there is no mention of value. Since cockfighting chickens provided a popular entertainment, they may well have been exchanged for a variety of merchandise.

Rent payments for the use of manor or estate lands were quoted in numbers of cattle, sheep, pigs, poultry and often, eggs. Capons were mentioned as payment for some of the rents. Poultry was used as a *shevage payment*, an approval fee to leave the manor. The following is an example:

Rental fee for 9.4 acres was rendered 57 hens and 443 eggs.

Often the rents would spell out specified numbers of capons, geese and ducks. Also included was an assortment of livestock, grain, wood for burning and money.

Rental of the estate owned by Earl Panure consisted of 1843 pounds money, 1586 botts of wheat, barley, oatmeal, 8 geese, 458 capons, 456 chickens and 312 hens.

The importance of poultry raising during the Charlmagne Reign, 1096-1270, is shown by regulations that were installed regarding the number of poultry and geese that farmers had to keep on their farm. These regulations were called Plumgraven.

In the year 1300 price controls were established in London by Mayor Elias Russell, by consent of the King and nobility, and authorized by an Act of the Common Council, which regulated poultry prices.

Eggs were used as a medium of exchange. *(Little girls went to the market with eggs in a basket to barter for tarts or ribbons for their hair.)* It was stipulated that no huckster was permitted to go out of the city to meet those bringing in poultry. The purchase arrangements were to be confined to the city. The time of day for buying and selling and a designated place for

Perhaps there may have been earlier coins than the one produced in 700 B.C., from the temple of Artemus, at Ephesus, but we know of none earlier. Indeed, the chicken has been honored much through history, playing an important part in the lives of many as well as the food economy

HIMERA.
B.C. 481–472.

ASIA.
B.C. 480.

ASIA.
B.C. 480.

CARISTUS.
Obverse—Cow suckling Calf.
B.C. 400–300.

ASIA.
B.C. 700-480.

HIMERA.
Reverse—Crab.
B.C. 400-300.

the bargaining was made known.

In 1652, the Common Council in London would fine an individual found guilty of feeding poultry, goats, hogs or cattle in the streets, three shillings, and four pence for each offense.

Not just anyone could deal in poultry in 1752. In England, the King and Queen granted a *Poulters Company* (license) to do business. An *action* (loss suit) was initiated by a company of poulters against a poulterer who sold poultry without serving the stipulated years as an apprentice.

Red Pyle Modern Game Bantams

Stamps
The world over,
honor the chicken

K10
ZAMBIA
REPUBLIC OF KOREA
30
50
DDR
DDR
20
POSTA
3.20 LEI
R.P. ROMINA
1981
CUBA
CORREOS
PINTO
GALLOS DE LIDIA
13
1981
CUBA
CORREOS
CANELO
GALLOS DE LIDIA
1
POSTA
1.35 LEI
R.P. ROMINA
55 BANI
POSTA
R.P. ROMINA
30 BANI
POSTA
R.P. ROMINA
1 LEU
POSTA
R.P. ROMINA
40 BANI
POSTA
R.P. ROMINA

Chinese Pre-incubated Eggs

The *Encyclopedia of China* states that this country shipped pre-incubated duck eggs. That inspired me to write the Poultry Science departments of many agricultural colleges, but I didn't receive any response on this unusual procedure.

I wrote to my friend, Ching Chang Wang, in Shanghai, who said that the eggs are shipped four to five days in advance of the hatching time. I also contacted Dr. Willard Hollander, a retired geneticist from Iowa State University. He was the only person who was aware of this practice and informed me that for years the Chinese have been shipping duck eggs from Taiwan to be hatched in Hong Kong.

I learned the International Crane Foundation in Wisconsin has shipped pre-incubated crane eggs to Russia. These eggs were placed in a mobile incubator and because the shipment was so valuable, an escort was sent along.

Incubation Since Ancient Times

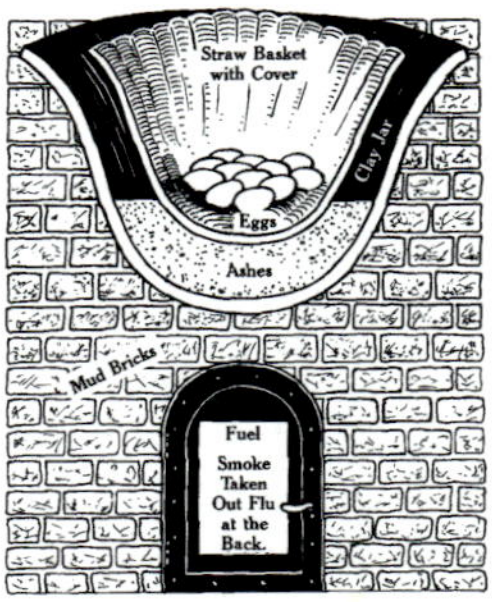

I'm not really sure when the importance of hatching chickens finally dawned on early chicken breeders. It was reported that the ancient Romans hatched eggs with heated stones but there is doubt about the extent of this hatching. Very early, the Chinese and Egyptians began artificial incubation. Aldrovandi, Italy, wrote precise instructions for making an incubator, very crude by our standards, but clever for ancient times. The following is an excerpt from his book, *Aldrovandi on Chickens* and outlines his incubator.

To hatch more eggs provide a container 4 foot high. There are three levels, each with a door. Each level, a floor would have a hole in the center for heat to rise up from the heat source at the bottom. The first and second levels to be made of small slats or woven twigs. Saw dust on each floor level. A funnel above the lantern to channel the warm air to the above three levels. Check eggs after four days for infertile, holding them up to the light. Infertile eggs to be removed. Check temperature of the eggs by holding them to the eye lid. Eggs to be turned daily.

Hatching egg in bosom per Aldrovandi

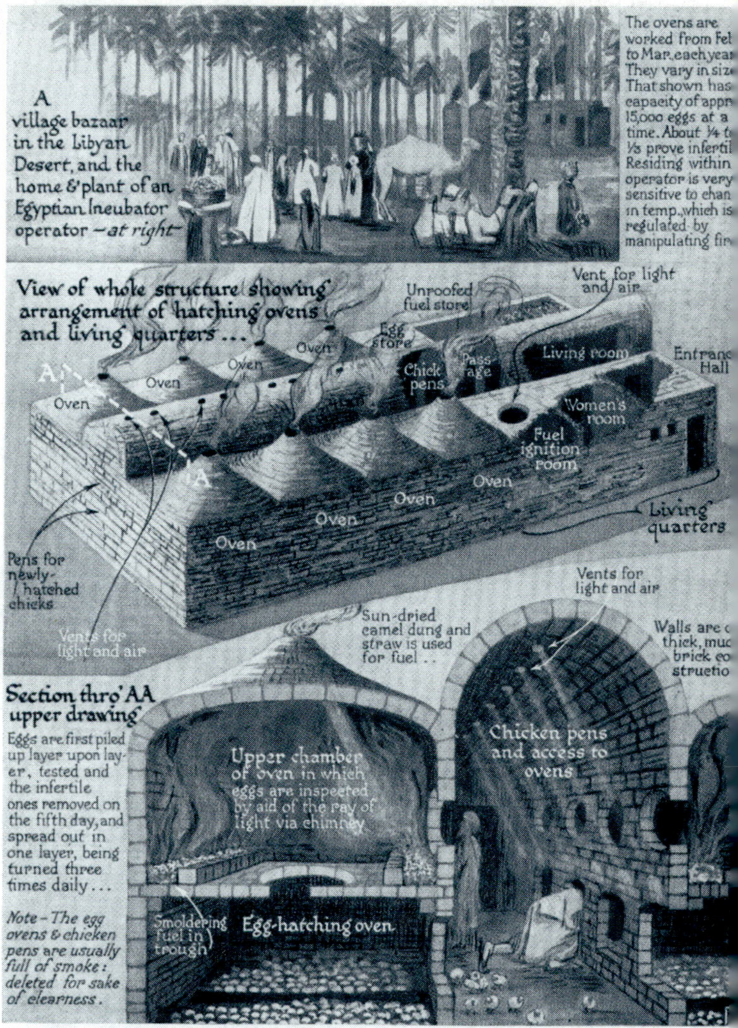

Drawing by Charles E. Riddiford

A PICTORIAL DIAGRAM OF THE STRANGE AND INGENIOUS ARRANGEMENTS OF THE GREAT EGYPTIAN INCUBATORS

These hatching ovens have been in use in Egypt for many centuries and are much the same to-day as in the time of Moses. There are several hundred such establishments in the Land of the Nile, producing from 15,000,000 to 20,000,000 chicks a season. The operator usually lives in the incubator. He has no thermometer to help him regulate the temperature, but by the "feel" of the air recognizes when the fires need attention.

Drawing by Charles E. Riddiford. Courtesy of the National Geographic Magazine

Professor Aldrovandi was aware of the oven incubators in Egypt. The French book, *The Discours Oeconomique* mentions hatching eggs with the heat from dung in a bed of feathers and to turn the eggs frequently to share the heat. This incubation may well be credited to the Chinese.

Egypt had the first large volume hatching of baby chicks, dating back to the time of Moses. The book from 1750 by M. de Reaumur, *The Art of Hatching and Bringing Up Domestic Fowl,* states that the Egyptians should be more renown for their oven incubators, known as Mamals, than they are for the Pyramids. National Geographic Magazine, April 1927, quotes an estimated 15,000 to 20,000 chicks a year were produced in these crude incubators. The article goes on to state that in 1895, there were 150 Egyptian oven incubators. Each one was capable of hatching 300,000 chicks a year. They were 60 feet by 100 feet and the operators literally lived in these incubators. Eggs were hatched with a *toll* arrangement. The supplier of the eggs would get two chicks for each three eggs. The number hatched beyond the two chicks per three eggs was a hatching fee or considered profit.

Hatches of 75% to 80% is a testimony to the efficiency of the Egyptian incubators and some of the Mamal incubators are still in use today. However, the modern commercial poultry industry of Egypt is using automatic incubators that are more efficient because of labor costs.

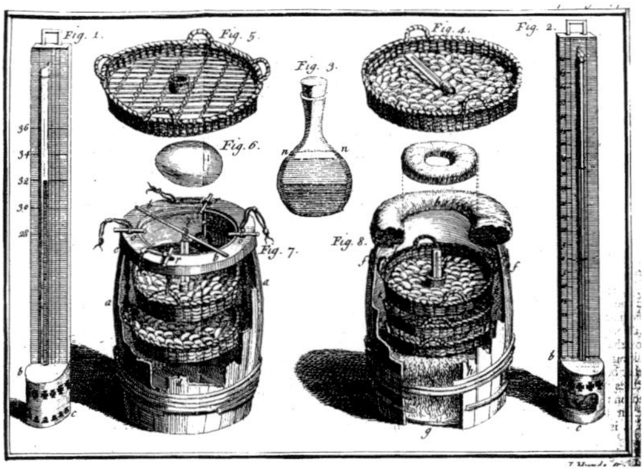

From the 1750 book by M. De Reaumur, inventor of the thermometer: The Art of Hatching and Bringing up Domestic Fowls, an early incubator.

French scientist, M. de Reaumur is credited with the invention of the first incubator though it was never commercially produced. In 1843, the first American patent for an incubator was granted to Napoleon Guerin, New York. In 1848, the Centelonian Hydro Incubator was introduced by Contello of the Modern Poultry Farm in Chiswick. Rouiller invented another incubator called the Legion.

Sir Edward Brown, England, reported at the

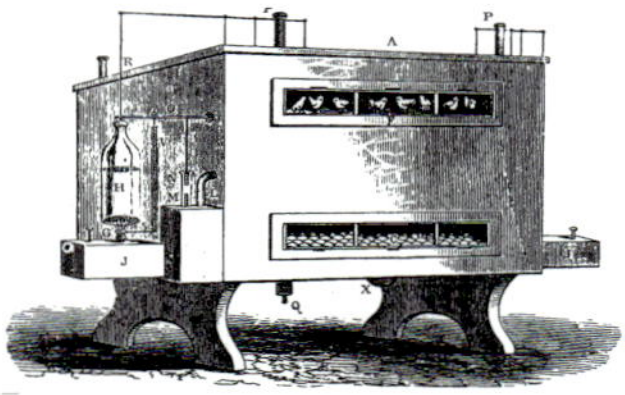

Ancient incubators make one appreciate our modern electric, automatic-turned incubators

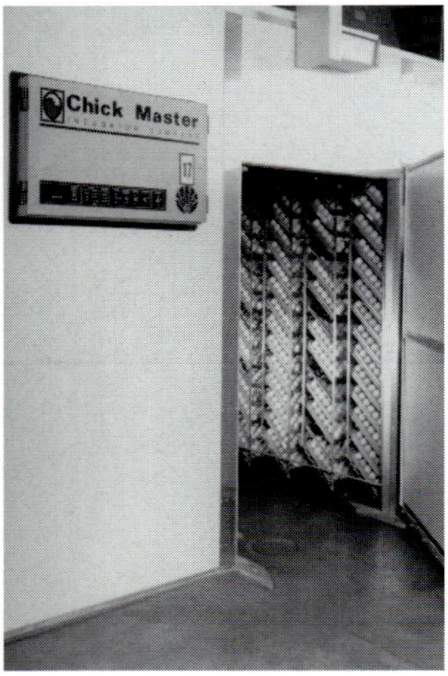

Modern incubator

International Exhibition in London in 1851, that an incubator being exhibited attracted considerable attention. His own experience with the Boyle incubator was in 1877.

The following compilation led to our modern incubator.

1878 Hydro incubator was introduced at the Paris Exhibition.

1884 Monarch incubator was manufactured by James Rankin.

1888 Prairie State incubator was advertised and a great success.

1896 A 20,000 egg capacity incubator was offered by Charles Cyphers Company.

1902 More than 200,000 incubators were exported from the United States.

1920 Buckeye Incubator Company offered the first cabinet incubator, kerosene lamp heated with electric fans.

1923 Petersime Incubator Company offered the first electric incubator.

1929 Bundy Incubator Company introduced the sepa rate hatcher. Smith Incubator Company offered the first walk-in, 47,000 egg capacity incuba tor, that was steam heated with electric fans. Petersime Incubator Company introduced a sim ilar incubator with electric heating and auto matic turning. Chickmaster offered a new incu bator plus conversion units for Smith incuba tors that converted them to electric heat and automatic turning units.

All over the world, the chick hatchery industry now benefits from the ultimate in scientific labor-saving features. Ultra-modern incubators provide efficient moisture and heat controls with excellent provision for perfect sanitation during the incubation period.

Egg Customs

Long ago, the Chinese dyed eggs to celebrate the return of Spring. The Egyptians, Persians, Greeks and Romans followed that tradition. But, to early Christians, eggs symbolized the mystery of the

Resurrection. They were consecrated and used in church services at Easter time.

It may have been a fourth century edict prohibiting consumption of eggs during Lent that started the practice of decorating eggs for Easter in Britain. Since hens couldn't be persuaded to stop laying for 40 days, eggs were hard-boiled and given to children as playthings for Easter.

In many parts of Britain, decorated eggs are still called by the ancient name of Pace or Pasch eggs, a name derived from the Hebrew word *Peseach* (Passover).

Easter eggs have long been associated with the Christian festival of Easter. The custom of giving eggs goes back to a pagan tradition. Some say that the English word, Easter is derived from the Saxon *oster* meaning *to rise*. It was an old Aryan belief that the sun rose on Easter Day and its appearance was greeted with dancing and celebration. The new sun was symbolized by red and gold eggs which were exchanged to commemorate the birth of Christ.

In Czechoslovakia and Hungary, girls colored and decorated eggs for their sweethearts. If a man received a number of eggs on Easter Day, he was considered a good catch. He would put the eggs away for a period of time to test the strength of the donor's love. The eggs whose coloring did not fade belonged to the girl he would marry. Girls could hedge their bets by giving eggs to more than one man. They had to wait for the men to collect the eggs they had decorated. A girl whom nobody favored, was left with her eggs and told to hatch them herself.

There is a legend that says the tradition of decorating eggs started with the birth of Christ. The Virgin Mary is said to have painted red, yellow and green eggs to delight the baby Jesus.

In Poland, when a woman gives a man an egg, it is a token of love.

In Hungary, when a gentleman is given a pair of eggs, he knows his sweetheart has transferred her affections...elsewhere

Eggs were believed to have magical powers to insure

Egg art
by Phyllis Burdett,
England
Gift to
Loyl Stromberg
1989

fertility. Slav peasants smeared their plows with a mixture of eggs and flour to obtain a good harvest.

In France, during the 17th century, the bride broke an egg as she entered her new home.

Carl Faberge, Switzerland, emigrated to Russia and is the most famous egg decorator in history. He produced many artistically created eggs for the Romanova family of Russia. Faberge also hand-crafted egg jewelry. Though his beautiful egg's started in Russia, they are known in eastern Europe including Hungary, Romania and Czechoslovakia. Faberge originals are priceless and most of them are in museums.

Sixteenth Century portrayal of the Egg Dance
Joan Galle

Culinary Facts

Boiled eggs for breakfast started as a tradition in China, Taiwan and Thailand. In these countries, eggs are boiled in natural hot springs. Enterprising hawkers offer eggs to anybody who neglected to bring their own. These naturally cooked eggs are thought to have a salubrious effect.

One theory about the origin of fried eggs is that a cannibal dropped an egg on a hot rock.

It was the Greeks who first discovered how to boil a chicken.

The French introduced scrambled eggs, a claim by Brillat Savarin, famous for his grand cuisine.

Imperial Rome's rich ate food that was sophisticated and elaborately prepared. *Lordly eggs warmed in their wisps of hay together with the hens that laid them.*

In the Dark Ages, a hen was valued more for her egg output than for her meat.

In ancient China, eggs were so scarce, they were pre served in caustic lime. This procedure is still used today. The inside of the preserved egg shrinks to a dark jelly-like material. These delicacies are called *thousand* or *hundred* year old eggs.

In the 1600's, a Franco-Italian cookbook was pub lished by La Varenne and featured 60 recipes for the humble egg.

A Boston cooking school cookbook, published in 1896, by Fanny Farmer, has an entire chapter of recipes showing different ways to cook eggs.

In Indonesia, eggs are preserved in a concentrated salt solution. When cooked, these eggs are very salty. American settlers preserved eggs in a similar manner and were able to keep them for months.

Eggs are highly valued for their functions in food preparation. Eggs leaven food, thicken, coat, bind, clarify, and add color, liquid and nutrients to many dishes. They are a most elegant and useful commod ity.

Food excavated from ancient tombs proved that Egyptians used eggs in their breads in the time before Christ.

The dish, Chicken a la King, was introduced when King Edward VII of England ordered a meal made according to his own recipe.

Fanciful Facts

Here are some delightful sayings, thoughts and meanings of words, with a few facts thrown in for good measure!

Bad egg...doubtful character.
Curate's egg...one that is good in parts.
Duck egg...score of naught, especially in cricket.
Electron egg...very stale (from the days of hustling when these eggs were saved to throw at unpopular speakers.
Golden egg...great profit from little effort.
Good egg...fine person.
Nuremburg egg...one of the first watches made (egg-shaped) made in Nuremburg, Germany.
Pasch egg...Easter egg, traditionally decorated.
Philosopher's egg...medicine which was supposed to be an antidote against the plague.

From Poultry Museum in Japan

Eggs were a symbol of fortune in the days of yore. Your luck was lost if you took them in or out of a house after dark. And, if you let three drops of white fall from an egg into water on New Year's Eve, you could tell from the shapes formed what your fortune in marriage would be.

Research indicated that chickens were imported into England for food in 1577. These were called, *Hampers of Hens,* which points to the value of poultry.

5¢ candy bar offered about 1935.
Merchandising program included cars with a 'chicken body' that was most attractive. Amazing but they did not have public acceptance and disappeared from the market

Chicken Art, Stromberg Library

Caponizing was reported by the early writer, Varro,(116-26 B.C.). Through the ages, capons were very popular and mentioned in a number of books including those by Shakespeare. In 1865, African explorer, Dr. David Livingstone, reported seeing fine, fat capons in Zambezi. Because of our modern broiler industry, capons have lost favor, but to a gourmet, a capon will always be a welcome treat.

The political promise, *A chicken in every pot on Sunday,* is not new. It dates back to 1553-1610 and was a quotation from King Henry IV of France. *I want there to be no peasant in my kingdom so poor that he is unable to have a chicken in his pot on Sundays.*

In Shakespeare's work, *As You Like It*, he writes, *Truly thou art damn'd; like an ill roasted egg all on one side.*

Mark Twain said, *Put all your eggs in one basket, and watch that basket.*

Books About Eggs

Eggs is a very interesting book by Linda Sonntag, packed full of pictures, facts and superstitions. Included is information on the egg dance, medical prescription eggs, cosmic eggs, magic eggs, egg rituals, the Humpty Dumpty story and the Aesop fable of the goose that laid the golden eggs. There are pictures of beautifully decorated eggs with an explanation of their significance when given as gifts for Easter. Many stories are presented including the one about the 1,000 year old Chinese eggs. Russian Ukraine eggs are included.

Two other books devoted exclusively to eggs are *Egg Craft*, by Violet Schafer and *An Egg For Easter* by Venetia Newall.

Chapter Four

Historical Poultry Researchers

Sir Edward Brown

In presenting portions of history for some breeds I will often quote Sir Edward Brown. His book, *Races of Domestic Poultry*, was published in 1906 and is considered a classic. He was given a full page of credits in the rare book, *Poultry Firsts*, by Lenora Hering, who considered him to be one of the first poultry researchers in the field of husbandry and poultry in general.

Sir Edward Brown was instrumental in the establishment of the *Utility Poultry Society* of England in 1896. This group became the World Poultry Science Association which was established in 1911.

The first World Poultry Congress was held in 1921, in Holland, with 16 countries participating. Brown was elected the first president of the World Poultry Science Association and held this position for the first three events. He served as Vice-President and was re-elected as Honorary President for the fifth term.

England honored him with Knighthood in 1930. During his lifetime, he wrote 21 books. Totally dedicated to his work, Brown traveled extensively gathering material and his vast correspondence has provided poultry fanciers with valuable research.

Darwin's Theory Questioned

In the years 1831 to 1836, Professor Charles Darwin traveled all over the world with a British surveying expedition. He was a naturalist and well-educated. Darwin became famous for his theory on evolution. In 1859, he wrote, *Origin of Species by Natural Selection*, which was about plants and animals. In 1868, Darwin's book, *The Variation of Animals and Plants Under Domestication* was published. In this book,

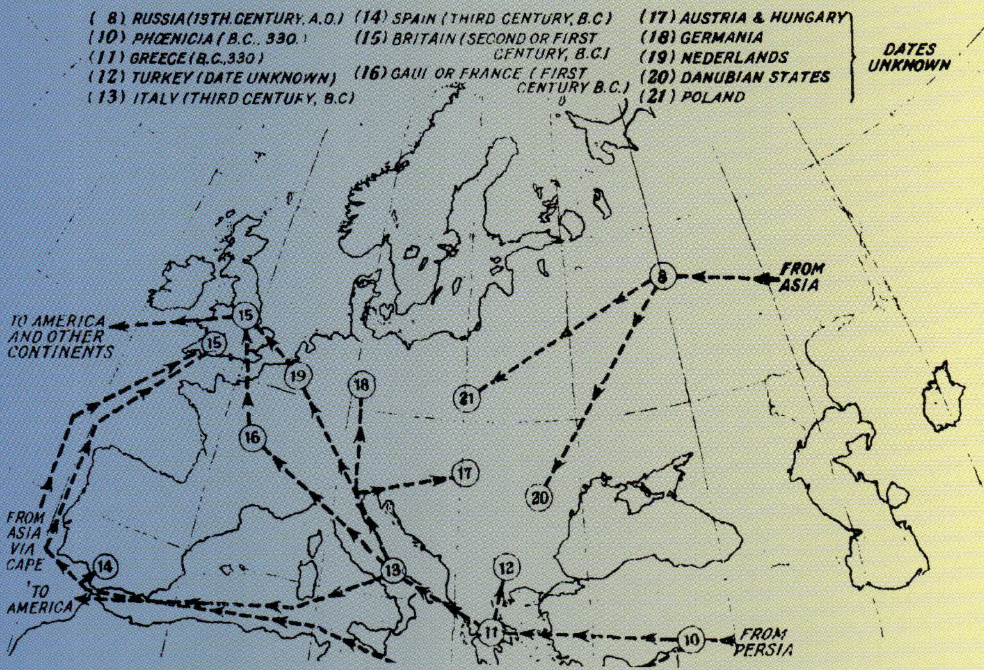

(8) RUSSIA (13TH. CENTURY. A.D.)
(10) PHŒNICIA (B.C., 330.)
(11) GREECE (B.C., 330)
(12) TURKEY (DATE UNKNOWN)
(13) ITALY (THIRD CENTURY, B.C)
(14) SPAIN (THIRD CENTURY, B.C)
(15) BRITAIN (SECOND OR FIRST CENTURY, B.C.)
(16) GAUL OR FRANCE (FIRST CENTURY B.C.)
(17) AUSTRIA & HUNGARY
(18) GERMANIA
(19) NEDERLANDS
(20) DANUBIAN STATES
(21) POLAND
DATES UNKNOWN
FROM ASIA
TO AMERICA AND OTHER CONTINENTS
FROM ASIA VIA CAPE
TO AMERICA
FROM PERSIA

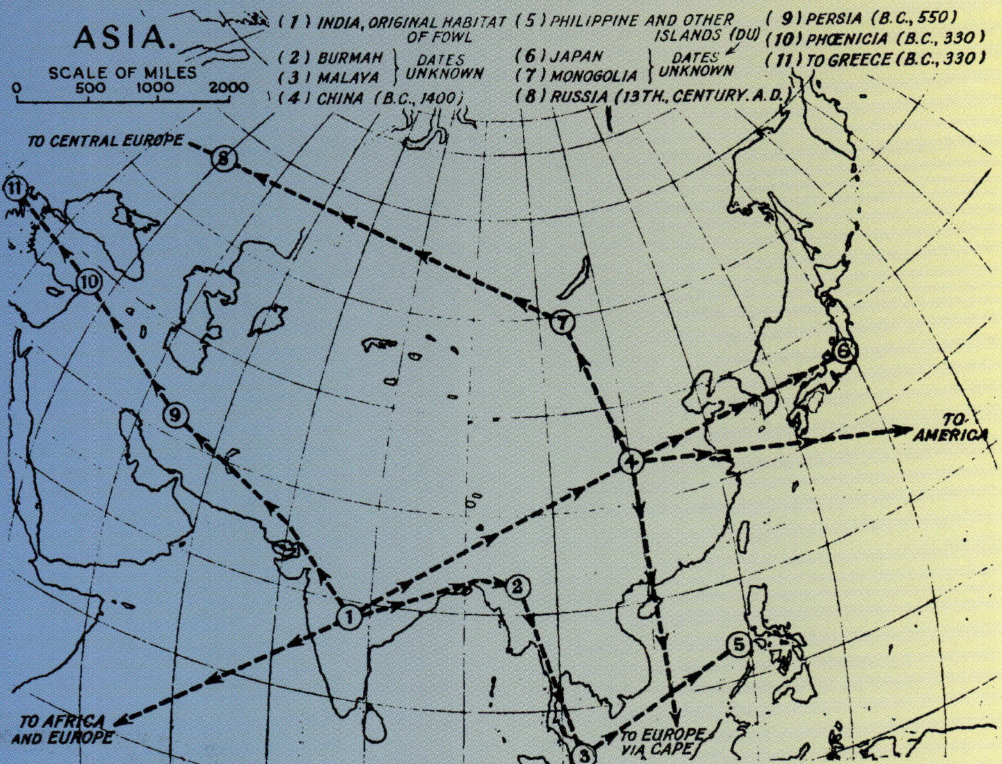

ASIA.
SCALE OF MILES
0 500 1000 2000
(1) INDIA, ORIGINAL HABITAT OF FOWL
(2) BURMAH
(3) MALAYA
DATES UNKNOWN
(4) CHINA (B.C., 1400)
(5) PHILIPPINE AND OTHER ISLANDS (DU)
(6) JAPAN
(7) MONOGOLIA
DATES UNKNOWN
(8) RUSSIA (13TH., CENTURY. A.D.)
(9) PERSIA (B.C., 550)
(10) PHŒNICIA (B.C., 330)
(11) TO GREECE (B.C., 330)
TO CENTRAL EUROPE
TO AMERICA
TO AFRICA AND EUROPE
TO EUROPE VIA CAPE

Professor Darwin writes about his theory of evolution of chickens, which was that present day chickens evolved from the Red Jungle Fowl, though he does mention five species of Jungle Fowl. My conclusion is that our chickens came from all species of Jungle Fowl, not just the Red Jungle Fowl.

Over the years, some would question the Darwin theory pertaining to Malay, Aseels, Cochins, Brahmas and Langshans. Sir Edward Brown, in his 1906 book, *Domestic Races of Poultry* was one of the first, W. B. Tegetmeir, Professor Jull, and noted geneticist, Dr. F.B. Huitt, *Genetics of the Fowl*, all questioned Darwin's theory. Dr. Huitt agreed that most breeds had their origin from the species of the Jungle Fowl, but states *some other ancestor, now extinct, gave rise to the Asiatic breeds.*

Dr. William Plant, Australia, author of *The Origin, Evolution, History and Distribution of Domestic Fowl,* does not agree totally with the Darwin theory. Dr. Plant's thinking is much along the same line as expressed by Dr. Huitt. I feel that the Malay, Aseels, Cochins, and Langshans did not evolve from the Jungle Fowls. However, the Darwin theory does apply to most

'To the Pastures'
New England

of our breeds, with only these exceptions.

First Poultry Shows

Teaching the entry to pose

England held local poultry shows between 1805 and 1840, they were not officially recognized shows. This was before the day of railroads and shows were held within a horseback ride of each other. In the north of England, Lancastershire and Yorkshire counties, the Bolton Greys, Bay Moonies and Pheasant Fowl were shown. This was before the Orpingtons and Cornish came into existence. Only hens were shown. The cocks were not permitted because shows were held in taverns and inns. They did however, have show specifications outlined in a book by Wingfield and Johnson, *The Poultry Book.* The prizes offered were useful household articles and highly regarded at that time.

In south England, there were poultry shows of a different nature. These featured fatted poultry. In that era, chickens were not bred for high egg production. Good meat chickens were very important to their economy, so south England had their own poultry show.

The first recognized poultry show in England was sponsored by the London Zoological Society in 1845 and was held at the London Zoo. Records show the following breeds were exhibited, White, Speckled or Grey Dorkings, Surrey Fowls, Old Sussex or Kent Fowls, Gold and Silver Hamburgs, Spanish, Malay, Gold, Silver Spangled, Black and White Polish. Bantam species were represented by Gold and Silver Spangled, and Black and White. It is significant to note that 1800 Cochins were exhibited in 1853 at the Birmingham Bingley Hall Show. That was the largest number of Cochins ever shown in English Poultry shows.

Old force feeding chickens for the markets

The first poultry show in the United States was held November 15th and 16th, 1849, in the popular Public Garden in Boston, Massachusetts. The large, beautiful Cochins dominated the show. This event was given considerable recognition by the newspapers. The estimated attendance was 8,000 to 10,000 poultry lovers. Gentlemen were charged admission and ladies were admitted free.

Daniel Webster exhibited his Javas and Geese and many prominent people attended including Colonel Samuel Jaques of Twin Hills Farms and the Honorable George P. Burnham, an early breeder of Cochins and Brahmas and author of the book, *The History of Hen Fever*. There were 1423 birds shown by 219 exhibitors, with a few cross-bred chickens included. I think the cross-breeds might have been the start of the Rhode Island Reds and Plymouth Rocks. Neither of these was perfected until many years later.

Other breeds exhibited included, Shanghae, Cochin Chinas, Malays, Javas, Wild Indians, Plymouth Rocks, Yankee Games, Fawn Colored Dorkings, White Dorkings, English Ravens, Bavarians, Polands, Bolton Greys, Dominques, Spangled Hamburgs, Italians, Guelderlands, Chinese Frizzled, White Spanish, African Bantams, White Cuba Bantams, Imported Swans, Pea Fowls, Guineas, Pigeons, Wild Geese, Domesticated Geese, Golden Pheasants, Aylesbury Ducks, Muscovy Ducks and Wild Turkeys.

From Paris
Exhibited about 1900, this unique method for feeding 400 hens in an hour to fatten in 18 days. Visitors could buy a fresh killed hen to take home for dinner.

Chapter Five

Fighting Cocks and Geese

Three Worlds of Poultry

The world of poultry can be divided into three groups, the first being that of the poultry fancier. Fanciers raise and breed their birds for exhibition at poultry shows and more importantly, for their own personal enjoyment. A poultry fancier makes personal choices, whether to have large fowl, Bantams, ducks, geese or turkeys. Many hobbyists enjoy raising several varieties including guineas, which though popular, are rarely exhibited.

World two is the commercial world of poultry which is concerned with producing high quality eggs and chickens for wholesale and retail markets. This is a highly competitive arena. Efficiency in production is the key to providing the public with top quality poultry and eggs at the most reasonable prices.

World three is the world of cockfighting which is probably the least known aspect of the poultry world, yet shines with history and enjoyment for the enthusiast. This is the world's oldest and most popular observation sport.

Cockfighting is the national sport of the Philippines. Many years ago, it was the national sport of the United States, and it is still popular in Mexico, South America and Puerto Rico. In Puerto Rico alone, statistics indicate 145 cockfighting pits, 588 judges with 216,000 cocks fought and $18,000,000 in total bets each year.

Tribesman of Borneo with Fighting Cock

Cockfighting being enjoyed in the 1800's

In India, cockfighting was established as early as 2000 B.C. There were ceremonial rites where chickens were foremost in their rituals.

Dr. Clifton Bryant, Virginia Polytechnic Institute and University, states in his article on the *Origin of Cockfighting*, that it may have started in 4000 or 5000 B.C.. Sir Edward Brown in his 1929 book, *Poultry Breeding and Production*, writes that the first mention of cockfighting was in the reign of Croesuc, King of Lydia.

The cockfighting chicken was considered as the national emblem of the United States, but lost by one vote in favor of the Bald Eagle. The wild turkey was considered as well, but did not fare well in the debate.

President Andrew Jackson was known to have kept fighting cocks in the White House stables, employing a man by the name of Jack Freer, to feed and train them. George Washington and his friend, Lord Halifax, were ardent cockers, though some historians believe Washington may have attended the fights but did not breed them. Abraham Lincoln loved cockfighting and it's recorded that his name of *Honest Abe* came from being an honest, skillful and impartial judge in refereeing cock fights. Benjamin Franklin was known also to be a popular cockfight referee.

Stone carving on walls of Bayon during reign of King Jayavarman VII

P.D. Selvrarj
and his fighter bird

Opposite
Cockfighting in Indonesia

Game cocks sparing,
drawing from 1817

Mosaic taken from the
ruins of Pompei
(destroyed 78 A.D.)
Now at the National
Museum, Naples

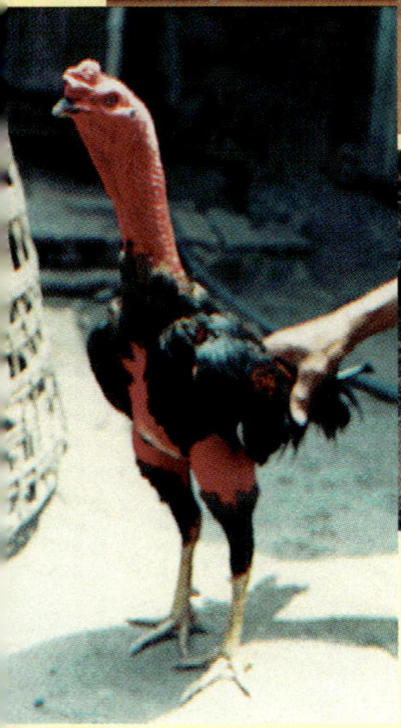

Above
Vietnames Game
in action
Bangkok/Vietnam

Below
Ancient enjoyment of
cockfighting in Japan

Today, in the United States, the sport of cockfighting is legal in only a few states. Still, it continues to be popular in many states, with the fights held secretly, evidenced by four specialty cockfighting magazines, two of which have a circulation of over 10,000 copies monthly.

Australia is know as the country first inhabited by convicts that were deported from England in 1788. The first fleet of convict ships included a few whose crime was stealing game cocks. Their punishment for stealing was a sentence of *7 years in Australia...with no return ticket.*

Through the ages the sport of cockfighting was enjoyed by many public leaders such as;

Themistocles
(Greece, 500 B.C.)
Alexander the Great
King Henry VIII
Sir Walter Raleigh
Abraham Lincoln
George Washington
Andrew Jackson
Thomas Jefferson
Henry Clay
John Calhoun
Martin Van Buren
Robert E. Lee

Fighting Geese of Russia

Sir Edward Brown refers to the Russian fighting geese, Tula and Arsamas, with information taken from an 1832 book by Bonnington Moubray, *Practical Treatise of Domestic Poultry.* It states that outside of Petersburg, they had no cockpits but did have goose fighting pits. It was a popular as well as profitable sport. The best fighters would command as much as 500 rubles, a very handsome price in those days. The geese were just as keen for battle as the fighting cocks. Though it was popular, the sport of goose fighting was banned for a period in the early 1900's, however, there is evidence that it continued on the sly.

It was my good fortune to have the help of Allan

Mustard, an Agricultural Attaché in the American Embassy in Moscow. He saw a news item in the March 24, 1987 newspaper, *Sovietskia Russia* about their fighting geese under the column *Old Russian Entertainment.* The following excerpt is a translation of the Russian article.

Every year, on the second Sunday in March, regardless of the weather, in the village of Orlovskia in southern Russia, and from surrounding neighboring villages, gather the fighting goose fanciers. You can

hear the lusty goose gaggle emanating from the large baskets in which they are brought to the small frozen pond. They sprinkle a little snow on it, it's too slippery.

After a signal, a pair of fighting geese are released on to the ice. They are surrounded by other geese, mostly females, who noisily escort the fighters as if to urge them to begin. The first pair are the younger geese, two years of age. They are divided into categories as in boxing, then the three year old, and four - they go up to five years.

To prepare geese for fighting requires special knowledge. In the summer, during the molting, they are fed just water soaked bread. They do not want them to grow too much down, only strong feathers. After that, until Fall, grass, bread and oats. From October until the end of February only dried and parched oats to strengthen the feather and body. Two weeks before the fights, Cookies and Prianiky -- to sharpen the wits.

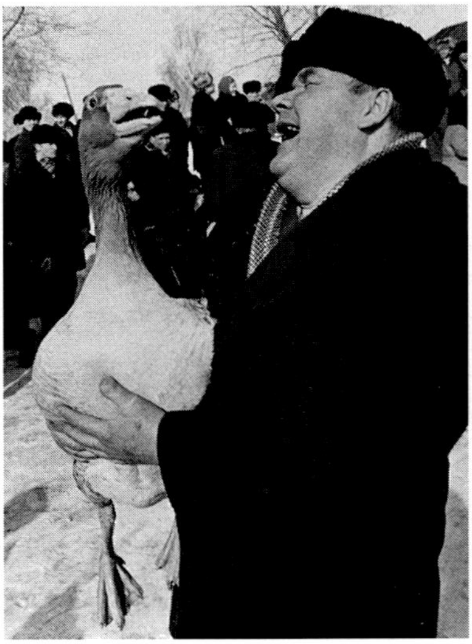

During the heat of the fight, beaks and wings are employed. At the shoulder of the wing the geese develop a large apple sized fist of bone and muscle. The strike is very damaging to the opposing goose with that fist.

There are certain rules to be observed. Geese can't use the beak on the head, legs or wishbones of the contestant. Infraction of the rules eliminates them from the contest. They are allowed only to hit the wings. Goose fighting is a very ancient tradition, at least 500 years old.

Professor Valimar Fisinin, Russia, reports that goose fighting is now a respected sport. It's very well know in the Gorky district, formerly Nishij Novgorod. Goose fights are held during the Shrovetide and though it was banned in the early 1900's, it still continues.

The most popular geese are the Tula. Sir Edward Brown's book of 1906, states there are three known breeds of fighting geese. The Tula are grey and range in size from 13 1/2 to 17 pounds and are bred in Tula, Kalonga, Tamboff, Kizan, Koursk, Valdimir, Nijni Novgorod, including the upper borders of Oka and upper parts of the Volga river.

The Arsamas are white and weigh 17 to 22 pounds. The Kholmogory are the largest, weighing 20 to 23 pounds, with longer, larger necks, making them stand out more than the other breeds. None of the fighting geese are recognized for their eating qualities, the taste of the flesh is more like wild game and they do not fatten as well as the Emden or Toulouse geese.

'Standing Pintail'
by
Geoffrey Dashwood
English, Contemporay
The Sladmore Gallery

Chapter Six

Long-Crowing Chickens of the World

There are six varieties of long-crowing chickens throughout the world. Three species are from Japan and one species each from Germany, Turkey and Brazil. These are very ancient breeds of unknown origin with limited historical documentation. They are not considered to be popular breeds, with the exception of Germany where they are exhibited at shows.

Japan

An article by F.S. VanGink appeared in a German poultry publication in 1956, and made reference to a Chinese breed of long-crowing chickens. It stated that old Chinese literature indicated this chicken existed in the Changkuo district of China between 206 B.C. and A.D. 1272 and was described as a fighter with long hackle and saddle feathers.

One breed became known as the Japanese Shokuko (Changkuo) and was introduced into Japan from China between A.D. 794 and A.D. 1186. Through the years the Japanese crossed this chicken with other breeds to produce long-feathered, long-crowing varieties.

Entries in contests are judged on the length of the crow, quality of voice, rhythm, and balance of high and low notes. Most birds crow from 12 to 20 seconds. Those who crow less than seven seconds are disqualified. Exceptional specimens have been known to crow for more than 20 seconds. Another requirement of exhibiting is that the sickle feathers of two-year-old

males must be a minimum of 70 centimeters in length.

The best known Japanese long-crowing chicken is the Toutenkou. It was named for its vocalized crow that sounds like *Too-ten-kou.* First bred in the Kouchi Perfecture region, it comes from the area where the Ongradori-type of chicken also originated. The average weight of the adult male is 2250 grams, the hen and cockerel, 1700 grams, and the pullet, 1350 grams. The Koeyoshi has a history of 350 to 400 years and was originally developed in Malaysia or Indonesia. These chickens act as musical alarm clocks, crowing three times a day. In the course of their vocalization, they cover a range of three octaves. The crow starts as *Cokkoogoo-oo* for two seconds, then stops for three seconds and continues for seven more seconds. In addition to its long crow, the females have a reputation as fairly good egg layers and the males are known for their cockfighting ability. Their weight averages 4.5 kilograms.

The Tomaru has been known since 1870. It has an impressively long tail as well as a long crow which typically lasts from seven to ten seconds but can stretch to 23 seconds. Avid breeders of Japanese long-crowing chickens hold crowing contests, but it has not been determined that they are public affairs.

Opposite
Toutenkou of Japan

Above
Koeyoshi of Japan
Toumaru of Japan

Bergishe crower of Germany

Germany

The Bergse Kraaier is known by two other names, Old National Breed of Germany and *Mountain Long Crowers.* Their history is not precise, but the belief is that Spanish refugee monks brought these large fowl with them to the village of Wuppertal in the Bergische land of Germany. In Westfall, there was crossing with local breeds and these chickens became known as Straw Hens.

Research indicates that the Bergse Kraaier appeared in different areas of Europe. Sir Edward Brown, wrote that this breed resided in the hill district of Rhensh, Prussia. He quoted M. Louis Vander Snicht of Germany, stating that the Bergse are from a sport of a common race of egg producers known as the Schlotterkamm.

The greatest interest in these long-crowing chickens came from the Bergische land in Germany near the cities of Solingen, Elberfeld, Barmen, Hagen, Mettman and Ludenscheid. Crowing contests are quite popular in Germany, with considerable public interest. Emphasis has always been on long-crowing ability rather than color or feather pattern. These chickens can crow four to five times longer than the average rooster.

Of notable mention is the manner in which these

singing cocks vocalize. They start with a very deep base tone that slowly becomes a little higher before it returns to a low tone. The bird ends the crow with the beak closed and the head near the ground. After the crow is finished, the breathing of the bird is a bit noisy.

Crowing contests are very popular in Germany, with considerable public interest. Competitors vie for cash prizes and value their winning chickens. At one time, a champion Bergse Kraaier cock could be traded for a horse or cow. Females are fairly good egg layers and weigh an average of 3.5 kg. Males tip the scales at 4.5 kg. They are bred in only one color, Golden. The Bergse appears as exhibition fowl in the *German Poultry Standard/Exhibitor's* book and photographs can be found in Horst Schmidt reference book, *Handbuch Der Nutz-Und Rassenhuhner.*

Mosaic of cockerel
Glasgow, Scotland
Museum

Turkey

Through the ages the people of the Denizili region of Turkey have prized the breed of long-crowing chickens that have their name. There are chicken statues in the city of Denizili that honor them. Although their exact origin is not known, the Denizili chickens are now propagated and distributed to the farmers by the Turkish National Agricultural Institute. At this facility, the birds are housed in pens and mated five females to each male. The females average 2 to 2 1/2 kilograms and the males run 3 to 3 1/2 kilograms.

The Denizili is not selected for color or type, but for the length and quality of their crowing. Short-crow cocks are not bred unless they are very melodious. There is no organized crowing contest, but there is a great deal of evaluation associated with crowing. Classification of crowing is extensive with not only length and quality being of importance, but also range, clarity, type and stance. Ratings can designate calls to be soprano, baritone or bass, shrill or satirical, true or false, brave, lion, wolf or bowed. Crowing ability in the first year can range 20 to 25 seconds, with the record length of the crow being 45 seconds.

Crows are of two types, false or true. The false crow has breaks. The true crow is performed without breaking and to its maximum capability. The four types of true crows are also based on the rooster's posture.

These are the categories of stance;

1 *Brave:* Normal stance, fluffed breast and forward extended neck.

2 *Lion:* Head and neck held high like a horse or bowing with feathers fluffed like a lion's mane.

3 *Wolf:* Head and neck fully extended with tail feathers touching ground.

4 *Bowed:* Head close to ground and thrust forward. This is considered an undesirable posture.

The beginning and end of the vocalization is the most important, with the crowing beginning slowly, building, then ending gradually rather than abruptly. The length is not as important as the beauty of the beginning and ending.

Musician fowl of Brazil

Brazil

There is one dominate long-crowing chicken in South America, the Musician of Brazil. It is sometimes called the Singer. There has been speculation that it was imported to Brazil from Europe and is related to the German Bergse Kraaier fowl, but that does not appear logical because the two breeds are quite different. While a few Japanese Toutenkou breeds are found in Brazil, the Bergse Kraaier is non-existent.

The usual length for the crowing of an ordinary rooster is two to three seconds. The Musician is eight to ten seconds. Length of crow is the first consideration of Brazilian breeders, while color patterns are unimportant. In appearance, most have a Single Comb and are similar in coloration to the Mille Fleur Bantam. Temperament is quite different for females and males. The females are very tame, docile and easy to handle, while the males are considered savage, and will attack people.

Documentation regarding the extent crowing contests in Brazil is scarce. It appears that it varies from within the country with some areas having none and others having a significant number of competitive displays.

Chicken shaped vessels in Horim Museum in South Korea (6th or 7th Century A.D.)

German Painting

Chapter Seven

Poultry Data

Hybrid Chickens

Commercial hybrid chicks are removed from the fanciers of exhibition poultry and the world of cockfighting chickens. Scientifically, they are truly a great accomplishment because of their efficiency. The first commercial hybrid was perfected by the Pioneer Seed Corn Company, Iowa. Their first chicks were sold in 1942 and merchandised under the name, *Hyline.* They were sold by an organization of farmers who became salespeople. The DeKalb Seed Corn Company, Illinois, followed with their development of hybrid chicks in 1945. They, like others entering the new era of the poultry industry, offered various lines, often identified by numbers such as Dekalb 289 and DeKalb 307. Selling hybrid chicks involved trapnesting, record keeping and scientific selection.

For many years, the need for eggs and poultry meat was supplied by farmers. The early 1920's heralded the start of commercial hatcheries which supplied farmers with their baby chicks. It was during that period that my parents became pioneers in the fast growing hatchery industry. They had been grain and cattle farmers, but built a modern chick hatchery in Fort Dodge, Iowa. Their hatchery increased in size to become one of the larger hatcheries in the United States. At peak production they hatched up to 2 1/2 million chicks and 250,000 turkey poults a year.

The demand for baby chicks grew in the United States to provide both eggs and meat. Between 1940 and 1960, there was a total of 10,000 hatcheries in the US. With the development of hybrid chicks, came the advent of large commercial egg factories. These corporations replaced eggs formerly supplied by farmers because they had the advantage producing a better, more uniform egg at a lower cost per dozen.

Opposite
Denizili of Turkey

Poultry Tips

Pinioning

Pinioning must be done on one wing only and with day old chicks, it is simple to do. Use very sharp scissors to cut the wing off at the first joint, removing the last section of the wing. I recommend this be done in the evening so if there is any bleeding, it can heal over night.

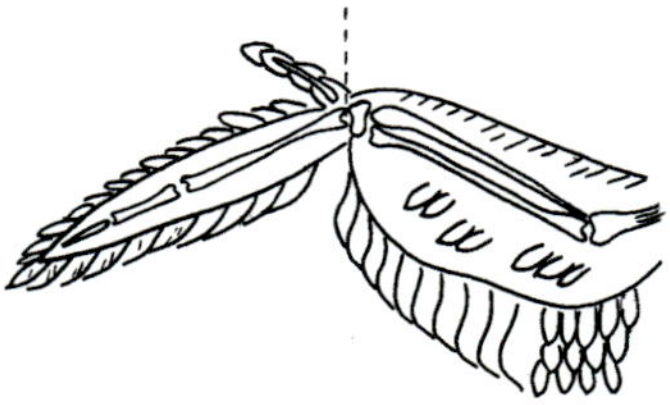

Caution: This has to be done on the first day to minimize bleeding. Older chicks will bleed more. Older fowl can be pinioned in a similar manner but you must use a very sharp knife. I have seen a sharp tinners' snip accomplish the purpose very nicely. Pinioning can also be done with a Lyons Electric Company electric debeaking machine which is both fast and effective.

To stop bleeding, dip the tip of the wing in tannic acid or liberally pat on baking soda. In the event the bleeding persists, you may have to tie string around the end

Various combs and feathers

of the wing, much like a tourniquet, to curtail the bleeding. If you are wary of doing the pinioning yourself, veterinarians will help you. Some poultry organizations may disqualify pinioned birds for exhibition.

Egg Eating

If you're having trouble with chickens eating eggs, it helps to debeak the flock though this may not be permissible when exhibiting. Provide plenty of nests and using the roll-a-way type can be helpful. Gather the eggs frequently. Most eggs are laid before noon every day.

You can try breaking open several eggs and mixing the contents with fresh, hot, red pepper. Pour the mixture back into a number of half egg shells and place in the nest where the eggs have been eaten. Offenders and innocent hens will soon learn that egg eating is very distasteful. Keep plenty of fresh water available and provide an oyster shell feeder for your flock. Make sure to keep the oyster shells clean.

Bumblefoot

For those not familiar with Bumble foot, it is a bruised chicken foot that has swollen to the point that the chicken limps and has difficulty walking. The swelling is obvious and the name *bumblefoot* is appropriate.

I have read that it can be corrected with a strong treatment of terramycin in the drinking water, but lancing seems to be a more positive solution. Lancing is accomplished with a good sharp knife to make an opening sufficiently large enough to squeeze out the *core* and accumulation of pus. The wound should be cleaned with a disinfecting solution and paper toweling. A final treatment of Iodine should be applied and the foot wrapped in gauze to keep it clean.

My observation is that bumblefoot is most often caused by birds jumping from high roosts onto a hard floor with little litter. I believe a good sanitation program with sufficient litter will help.

Feather Eating

I associate feather eating with a nutritional deficiency. But, this may not always be the case. I think it can happen as the result of over-crowding or it may be a nervous habit. Try soaking dry bread in vinegar and feeding it free choice.

Illness in Poultry

Being a member of a poultry organization can be a real plus. Often more experienced fanciers can help you with a sick chicken. If this help is not available, the Poultry Science Department of an Agricultural University should be able to answer your questions. I also recommend contacting the veterinarian of the nearest large zoo.

Because we live in an age of specialization most of your farm animals or small animal vets may not be able to help you. Your area feed dealer may be able to refer you to someone knowledgeable in poultry diseases.

Paul Chapman
Norwich, England

Poultry housing in Tanzania

Predators

Predators can cause sizable losses in small flocks. Not only is there a loss of birds but other chickens may feed on the carcasses making identification of the predator difficult. Dead birds may remain hidden and possibly result in further loss through botulism.

Predators may be identified by the signs they leave like tracks, droppings, hair or manner of feeding. The following feeding habits can indicate the kind of predator you may be dealing with.

Dogs: The feral dog kills wantonly. Carcasses are badly torn. Many birds are killed.
Fox: Usually takes one fowl and leaves premises trailing drops of blood and feathers. May partially bury chicken. Can climb fence of considerable height.
Opossum: Usually kills one chicken at a time, often mauling it. Eggs may be smashed.
Owl: Heads and necks missing. One or two dead birds.
Raccoon: Tends to eat heads off many chickens in one night.
Weasel, Mink: Bites through skull, back of neck or under wing of poultry. May eat only back of head and neck, kill several birds in one night and place neatly in a pile.

With valuable exhibition stock, poultry fanciers should give serious thought to protection against predators. It is amazing how wild animals have adapted to life within the cities. A well-constructed yard that includes over-head protection extending from your poultry building, can be provided with heavy netting or fencing using one inch openings. Keep in mind that predators can dig so your defense needs to go at least one foot into the ground. I have seen heavy plastic sheets offered for this purpose and they work well. Protecting against predators is expensive but so are your birds.

Decline in Egg Production

Under ordinary circumstances, the decline in egg production is about 20% of the preceding year. If the first year's egg production has been retarded by improper management or a disease, a hen may lay more the second year than the first one, if this has been corrected. Normally, a pullet that would lay 200 eggs during the first year can be expected to lay about 160 eggs the second, and 130 eggs the third year.

Poisons

There are many different poisons that can affect chickens. Plants commonly used for shade, such as oleander, castor bean, chinaberry and tamarisk are all poisonous to chickens. Chickens don't normally eat these plants, but shedding seeds frequently contaminate the feed. All of these plants, except tamarisk, have a definite toxin which will poison chickens, causing nervous symptoms and death. Tamarisk leaves or stems can cause imputation of the crop. Laxatives made from Epsom salts may help to correct the effects of the plants.

Salt poisoning is caused by chickens eating ice cream salt or brine thrown carelessly within their reach. Salt causes severe enteritis which is difficult to treat and the chickens usually die.

Botulism or limberneck is brought about by chickens eating spoiled feed or rotting meat. The organisms causing the poisoning are growing in the food. This poisoning is characterized by a paralysis of the neck muscles preventing the affected chickens from raising its head. Treatment must be administered

Greeting Card from Sweden

Araucanas
Painting
Mary Ellen Johnson

immediately to save the chicken. Epsom salts or other laxatives may help. The condition can be prevented by sanitation and careful feeding. Spoiled canned food, meat or vegetables should never be fed to chickens.

Removing Spurs

To remove a spur from a chicken, place a hot baked potato on the spur and hold it there for a few minutes. Remove the baked potato, twist the spur and you will find it comes right off. There is no blood or mess. Use another freshly baked hot potato for the other spur. This technique really works well.

Crowing Roosters

Some people keep a rooster so they can enjoy the cheery crow of their early morning greeting. By contrast, there are those, especially city dwellers, who hate the disturbance. Unfortunately, this dislike can make a poultry fancier an unpopular neighbor.

What can you do to avoid the problem? Here are a few ways to minimize the noise. Sound-proofing your building can be one solution, but this is rather expensive.

The design of the building, especially its ventilation system, will greatly reduce noise levels escaping from the structure.

Make ample use of landscaping and planting. Evergreens muffle noise. Keep the birds confined until mid-morning. Remember that Bantams tend to crow more frequently, but not as loudly as the large fowl. Voice characteristics tend to be inherited.

Keeping the lights on at night can help reduce the crowing to a degree. Playing a radio will assist in obliterating a piercing crow. With close neighbors, remember that what you are enjoying may well be a source of irritation to someone else.

If you live in a city, write Professor John Skinner, Poultry Science Department, University of Wisconsin, Madison, Wisconsin, and request his bulletin, *Fact Sheet: Raising Small Animals and Fowl in Urban Areas.* It provides very helpful information.

Devoicing Peafowl

The shrill cry of the peacock could be called a scream and is especially significant during breeding season. At other times, this fowl will voice its objection to an intrusion into their territory by a stray dog, wild animal, snake or visitor to their yard. To many people the cry of the peacock is a welcome reminder of their beauty and presence.

Opposite Bangladesh Poultry House

Below Morocan Chicken

The scream of the peacock in highly populated areas can make the devoicing of this fowl advisable. To help you, I am quoting the bulletin of Dr. W.B. Gross, Virginia Polytechnic Institute, regarding this operation. *(The following article is reproduced in its entirety with permission).* I must issue a word of caution. This operation is not for the novice and should be performed only by a surgically experienced veterinarian.

Operation for Devoicing Peacocks

Introduction

Birds produce sound by vibration of the external tympaniform membrane (ETM), with modification by the trachea, pharynx and mouth. The mechanism of sound production is similar in the peafowl and turkey. The syrinx is at the posterior end of the trachea, just dorsal to the heart, and attached to the ventral side of the esophagus. The external tympaniform membrane fills the narrow space between the tympanum and the first bronchial half-ring. At the syrinx the bronchi change from round tubes to much taller and narrower passages, with the ETM on the lateral side and the thickened internal tympaniform membrane (ITM) on the medial side.

To make sound, the syrinx is shortened by contraction of the sternotrachealis muscles. This pushes the ETMs into the air passages. Air is forced out of the bronchi and through the syrinx. Because of the high differential in pressure between the bronchi and trachea, air velocity is high and the membranes are pulled farther into the air passage (Bornelli effect), with vibration (and sound) resulting.

To prevent sound production it is necessary to stop the ETM from vibrating or to change the shape of the syrinx so that the ETM cannot be brought into close proximity to the ITM. This report describes a method for reducing the sound intensity of peafowl by altering the space of the syrinx.

Materials and Methods

The route through the thoracic inlet alongside the trachea to the syrinx is narrow, thus making it difficult to see the operative site or to manipulate instruments. A lateral view of the surgical
area shows the limited amount of space leading to the operative area. It was necessary to build both a viewing light and a cautery tool.

Special Equipment

A small battery-powered light was made by placing a grain-of-wheat lamp at the end of a piece of Teflon tubing about 25 cm long, with openings sealed with silicone rubber. A cautery tip was made from a 15-cm piece of insulated copper wire by hammering, filing, and insulating all but the tip with Plexiglas. A special ground line to the bird was made, ending as an 18- gauge needle which was inserted into the pectoral muscle. The cautery employed was an Electro-Scapel, Mark C, supplied by the Miller Surgical Company, Chicago, Illinois.

Surgical Technique

All directions are given from the observational position of the operator. After fasting for at least eight hours, the peafowl, weighing 2 Kg or more, is held on its with a special jig so that the thoracic inlet faces the operator. Aseptic technique is used throughout the procedures. Feathers are plucked or are held aside with clips to expose the skin over the thoracic inlet. A vertical incision is made through the skin over the thoracic inlet, and the subcutaneous muscle and crop are pushed aside by blunt dissection, exposing the intraclavicular air sac.

A hemostat is inserted into the air sac at a point nearest the keel. The opening is enlarged ventrally exposing the trachea leading to the syrinx. A light is passed down one side of the trachea to illuminate the syrinx, the white area at the end of the trachea just beyond the dark-reddish tympanum. The auricle is just above the syrinx, and many of the major vessels are lateral to syrinx. The cautery is passed parallel to the trachea with the tip pointed medially. Since the side of the syrinx is barely visible, great care should be made in placing the tip. The cautery tip is hooked over the distal edge of the tympanum, and the power of the cautery is carefully increased until it cuts the ETM.

The next cut is the most difficult. Cutting too much causes the peafowl difficulty in breathing, and cutting to little decreases vocal control. To reduce the possibility of cutting the auricle, push downward with the cautery tip. While cutting, draw the tip toward the operator and into the first cut. Be sure also to cut the internal tyumpani-form membrane medial to the ETM. Repeat the operation on the other side.

Suture the skin and subcutaneous muscle together while closing the opening so as to make an air-tight closure. The synthetic suture material is not removed. It is extremely important that no infection occur on the cut syrinx. Avian pus is solid and can fill the air passages, resulting in asphyxiation. Administered intramuscularly each day for three days is about 100,000 IU of penicillin and 0.125 g of streptomycin in an aqueous suspension. A single dose at the time of surgery is not enough.

Results

Over 150 such operations have been performed during the last 20 years. After about two months, healing is complete; about 80% of the birds had less than 1/10 of normal vocal intensity; about 8% lost less than one-half of the normal vocal intensity. A few birds with excellent control of intensity had difficulty in breathing. Most birds showed respiratory distress when exercised severely but not under normal conditions with three birds. The operation was impossible in three cases since access to the syrinx was so restricted that it could not be seen. The results from the operation are permanent, lasting at least 20 years. A few birds were lost in surgery because of hemorrhage into the respiratory passages. Most losses occurred while the technique was being developed.

'The Poultry Yard'
Jan Steen
1626 -1679

Discussion

Reduction of vocal intensity was accomplished by changing the shape of the bronchi at the tympaniform membrane from a long, narrow passage into a smaller, more rounded passage.

Other Bulletins of Interest

Removing the Vocal Cords of Fowl
Dr. A.J. Durant
University of Missouri
Columbia, MO
65205

Electro-Surgical Devoicing Poultry
H.P. Van Krey
F.X. Ogasawara
University of California
Department of Poultry Science
Davis, CA
95616

Devoicing the Chicken
Dr. W.B. Gross
College of Veterinary Medicine
Virginia Polytechnic Institute
Department of Veterinary
Science
Blacksburg, VA
24061

Carved chicken dish from Nigeria

A Method of Devoicing Fowl
David Tudor
Howard Woodward
Rutgers State University
New Brunswick, NJ
08903

The preceding information is intended to be educational and not for diagnosis, prescription or treatment of disease or any other health disorder of chickens. This information should not replace competent professional advise. In all cases, please seek the advise and assistance of your local avian veterinarian. The author and/or publisher are in no way liable for use or misuse of this information.

Exhibition

The following is the criteria for exhibiting poultry.

Concentrate on a single breed.
Buy the Standard of Perfection and study your breed.
Coop-train your birds before the show to tame and calm them.
Visit with as many exhibitors as possible and ask questions. Weigh all answers carefully.
Visually compare your birds to all others in the class.
Invite others to attend the show with you.
Be a gracious winner and a courteous loser.
Show friendly interest in other exhibitors entries
Join and support both local and national clubs.

Here are some things you should not do.

Overcrowd birds going to the show. Place more than three birds in a single compartment.
Let your birds monopolize your attention. Study the class.
Exhibit birds that are sick, poorly feathered, out of condition or with disqualification's.
Visit with judges while they are working. Save your questions for later.
Complain about the judging, unless there is an obvi ous mistake. Remember, it is only one person's opin ion.
Take one judge's opinion. Show your birds under several judges before altering your breeding program.
Raise too few replacements. You must have a large number of candidates from which to select your show entry.
Vaccinate within 30 days of the show.
Show birds with lice, scaly leg or injury.

Scientific Poultry Facts
By Dr. Williard Hollander

The pulse-rate of a chicken is 350 beats per minute.

A chickens temperature is the equivalent to a high fever for us.

Chickens lack sweat glands.

Chickens don't need vitamin C in their food because it is synthesized in the liver.

Chicken lungs are not sacs like ours, but passageways for the air between the bronchi and the air-sacs in their body.

Aflatoxin poisoning from grain was first proven in ducklings.

Sometimes extra large eggs are laid containing as many as three yolks or even a complete smaller egg with a shell.

Chickens are subject to 217 reported diseases, and 83 varieties of parasites and pests.

Pea Comb in chickens does not look anything like a pea, nor does Rose Comb look like a rose.

The biggest adult chicken ever reported weighed 21 pounds.

The smallest adult chicken to be reported weighed 11 ounces.

More than one sperm enters the egg at fertilization and the extras temporarily take part in early cleavage *(formation of cells for the embryo).*

In 1936, a chicken was reported that was naked-necked and feather-legged on the right side only. It was a freak but it was a special occurrence.

Unlike humans, the red blood cells of a chicken con tains a nucleus.

Crosses of chickens with species like peafowl, guinea fowl, pheasants and turkeys has been successfully produced by natural mating. Crosses with Quail have been done via artificial insemination. All hybrids are sterile.

In 1858, Darwin showed that a large crest *(as in the Polish breed)* is based on herniation of the skull, often with incomplete bony cover.

Hens need not be impregnated to lay eggs. The rooster is necessary only to fertilize the eggs.

It may take more than two days for a chick to break through the shell.

Dilute acid (acetic or hydrochloric) will wash off the cuticle surface of egg shells, meaning that duck and quail eggs, can be made very white.

Ducks can harbor an influenza virus which may inter act with a human flu virus to produce a more danger ous strain.

Chickens with four legs can result from partial twinning in the early incubation of the embryo.

Early research done on male hormones used Leghorns chicks as test subjects. The amount of comb growth was measured by rubbing a hormone on the comb. It was absorbed through the skin.

Chickens have no piebald color. Any patterns are com parable to those of the other domestic bird species such as Magpie pigeons and Pomeranian geese. Chickens fed diets lacking caretenoid (yellow) pig ments over a long period of time, produce eggs with colorless yolks. Diets that are rich in carotenoid pig ments can make yolks deep yellow or orange.

Comb size tends to be minimal when chickens have plenty of sunlight.

Chickens raised in dim light tend to develop the extra large combs.

The first demonstration of a cancer-causing virus was in 1911 by Professor Peyton Rous, United States, using a curcuma tumor from a Plymouth Rock chicken. The virus is now known as RSV (Rous Sarcoma Virus).

Waterfowl molt all flight feathers at one time, experi encing a flightless period. Chickens molt the flight feathers sequentially so that flight is not hampered.

Cocks having the Rose Comb gene homozygous (R R) have been found to be much less fertile than others.

The voice of the chicken is produced not by the larynx but by the syrinx, which is just above the heart.

An Ancona hen lived over five years, ate 420 pounds of processed feed, produced 634 eggs of which ten were soft shelled. She produced 813 pounds of slop py droppings and finished her life nearly skin and bones but with enough left to make 287 cans of chicken soup.

A chicken has two ovaries. The left one makes eggs and the right one is a standby and is used only if something goes wrong with the left.

A chicken can run nine miles and hour. A pig can run 11 miles an hour.

Music for Chickens

Professor Gadi Gvaryahu of Cornell University, determined that 18th century classical music played in the growing pens of broiler chickens, is conducive to more efficient weight gains in the birds. There is also less mortality. While soft music enhanced the chickens growth, loud music had the opposite effect.

Professor Gvaryahu's interest in this research began when he was a student at the Hebrew University in Israel. A poultryman became frustrated while trying to move his chickens from one end of a long building to the other. The chickens bunched together and refused to move. There was a radio playing beside them and the poultryman moved the radio to the opposite end. The chickens followed him. This experience inspired further study with significant results. Professor Gvaryahu's work is now published and available by writing to Cornell University, Poultry Science Department, Ithica, NY, 14850.

By playing music, the rate of gain was 2% more for the females and 3% more for the males, adding up to a $14,000.00 a year savings on a million bird broiler operation. For the entire broiler industry in the US, there is a potential savings of $57,000,000.00 in one year.

How Chickens Communicate

By Dr. Willard Hollander

Crowing	Challenge, exuberance.
Chattering	Light conversation about noth ing in particular, keeps the group aware of the location of others.
Clucking	Assembly call of hen to chicks. Also by cock, *Here's something good to eat.*
Cackling	Alarm call or a hen announcing she's laid an egg.
Prolonged whistle	*I see a hawk*, or *I'm really scared!*
Squawking	*Ouch, that hurts!* Screaming induced by being mishandled.
Singing	Worried, bothered or looking for food.
Cock's courting note	Struts around a hen, sort of chuckling.

The Social Order of Hens

By Professor A.M. Guhl, Kansas State College

The *order of peck*, a rigid hierarchy among all chickens is determined simply by the problem of who can peck whom.

The Boss Hen - The Order of Peck

The social standing of hens has long seemed utterly haphazard to human observers, their gatherings in the chicken coop being generally regarded as a confusion of unrelated squawks and scratches. But by watching the pecking and pushing of any group of chickens the scientists have discovered that hen society is one of the most rigid in the animal world.

Every hen in the flock has her own position. There are certain hens she can peck, others who can peck her with impunity. She must defer to her superiors at the feeding bowl or in the roosting box, but her inferiors she may, and does, shoulder away ruthlessly. It is for the enjoyment of these privileges that the peck order among hens is established.

Dozens of flocks of chickens have been studied at

Kansas State College by Dr. Alphaeus M. Guhl, who assembled eight white Leghorns for his research. In less than an hour every hen in the flock had recognized the superior pecking ability of hen number one, who thereafter enjoyed the prestige in the group of eating and roosting first. But for five days the other hens fought bitterly for a place high in the hierarchy, finally settling down in a specific order.

Hen number eight, who was pecked by all the others, had one compensation for her troubles. She was permitted at the feeding bowl by hen number one, but only because she was so far down the social scale that number one probably felt she could tolerate number eight with very little loss to her own prestige.

Social Upstarts Seldom Succeed

As top hen in the flock, number one had no psychological problems at all. She ate and slept when she chose and as a result became able to lay more eggs than most of her companions. Some of the others, however, developed alarming symptoms as a result of the constant frustration of having their toes pecked or enduring the haughty stares of their superiors.

Hen number two for example, who was infuriated at not being number one, had periods of storming around the yard and pecking every hen in sight. She even pecked number one a few times and was roundly thrashed for her temerity. Other hens, less aggressive in character, accepted their station in life philosophically.

From being forced away from the feeding bowl and the nesting box, the egg production of the lower classes fell off. But the less domineering qualities of these lower classes made them so attractive to roosters that their eggs had a better chance of becoming fertilized than those of the queen herself.

From time to time the social structure in Dr. Guhl's coops had tottered dangerously under the onslaughts of ambitious and embittered hens or with the arrival of a stranger. But it has never fallen, the stranger merely taking her place after a series of skirmishes to test where she belongs. The most dangerous moment comes when the hens grow very hungry. Then if a feed bowl is placed in the center of the coop, they ignore the social niceties and rush at it all together. But a peck or two and an angry glare soon settle that.

As an experiment, Dr. Guhl injected male hormones into the most plebeian hen of the whole flock, number eight became number one!!!

The above article appeared in Life Magazine, August 2, 1948 and is reproduced with permission.

I conclude that scientists found the chicken society is like a ladder. At the top is a despotic boss and on the bottom rung, a hen-pecked outcast. Humans seem to have an order of peck as well, *You can peck your wife's sister but you do not dare peck your mother-in-law.*

Ancient drawing of chicken raising England

Henny-Feathered Roosters

By Dr. Willard Hollander

Adult roosters of several breeds such as the Sebrights, Campines, and some species including the Guinea, have the same form and color pattern of plumage as the hens. In chickens, the explanation has been found. The evidence indicates that sensitivity to sex hormones is different in these breeds. This following is a list of facts relating to henny-feathered breeding;

Caponizing a henny-feathered cock results in change to cock feathering at the next molt.

Caponizing a normal cock results in no change.

Ovariectomy of any pullet will result in development of cock feathering at the next molt.

Injection of a female hormone into a normal cock while the feathers are growing, results in feminization of those feathers.

Injection of a male hormone into a normal cock while feathers are growing, gives no change. (Remains cock-feathered).

Injection of a male hormone into a henny-feathered cock while feathers are growing, gives no change. (Remains henny).

A cock-feathered capon of a henny-feathered breed when injected with a male hormone develops henny-feathers again.

This sounds a bit strange, but I can summarize by saying that the plumage of a henny-feathered breed responds to a male hormone as if it were the female hormone. This peculiar response is not shared by the comb or behavior. Crosses of henny-feather with normal breeds give sons showing moderate henny-feathering. Further breeding tests have indicated that a single partly-dominant gene is responsible, and has been given the symbol Hf.

Chapter Eight

Prominent People and Poultry

Queen Victoria received the first Cochins from China. She started the tradition of the Royal family in England to have an interest in poultry, attending the shows faithfully

Notable Poultry Fanciers

The following celebrities, royalty and statesmen enjoyed poultry.

Victor Borge, musical comedian, raised Cornish Games, Connecticut, and is credited with coining the name *Cornish Game Hens* and making them popular nationwide.

Senator Morris, Oregon, was a Bantam fancier.

Bill Cosby, entertainer, raised Bantams.

Kitty, on the television program *Gunsmoke*, raised Rhode Island Red Bantams.

Dupont family raised and exhibited fancy poultry.

Len Small, Governor of Illinois, raised exhibition poultry.

Queen Victoria, during 1840 to 1860, raised fancy poultry on an extensive scale. It was her interest that was responsible for the *Cochin Craze* and the start of the first English poultry show increasing the growth of poultry exhibitions and shows.

The Queen Mother of England was presented with a fine trio of Buff Orpingtons by the President of a poultry club. She had them exhibited at one of the leading poultry shows in London.

George Washington imported the first Chinese Geese and Golden Pheasants.

Daniel Webster exhibited Javas and geese at the first poultry show in Boston, November, 1849.

President Rutherford B. Hayes kept Light Brahmas in his White House poultry yard.

Thomas Jefferson raised Bantams.

Henry Ford enjoyed meditating in his chicken coop.

Robert Frost, poet, raised chickens.

Babe Ruth and *Dizzy Dean*, baseball players, raised Rhode Island Reds.

Rudyard Kipling, poet, was an early breeder of Light Sussex.

Colonel Garbisch, football player at West Point, exhibited White Rocks.

Senator John Chaffee, Secretary of the Navy and Governor of Rhode Island, gave his wife a trio of Rhode Island Red Bantams for Christmas.

President Lyndon Johnson, while a Senator, raised Silkie Bantams.
Colonel Sanders, of Kentucky Fried Chicken fame, raised Rhode Island Red Bantams.

Louritz Melchior, Danish opera singer, was a fancier of Cochins.

Clark Gable and *Carol Lombard* were chicken fanciers.

Pat O'Brien purchased Canadian Honkers from the Stromberg's. His wife raised Bantams.

Dick Powell, actor, raised and showed Silkies.

Greer Garson, actress, had White Cochin Bantams.

Lowell Davis, figurine artist, raised Cochins and showed White Leghorns and White Rocks.

Danny Kaye, actor, raised fighting chickens.

Ray Price, country music singer, raised fighting chickens.

Tom T. Hall, country music singer, had Bantams, large fowl and geese.

Raymond Burr, actor, enjoyed his chickens.

Dolly Parton and *Loretta Lynn*, country music singers, have chickens.

The following are known to have had chickens; *Lilly Pons* (soprano), *Marine General Harry Mad Dog Johnson, Walter Johnson* (baseball player), *Stu Irwin* (movie star), *Governor Alber Sharpy* and *Senator Lovegrove*, Connecticut, raised Belgian Bantams.

Above
The Queen Mother of England with Will Burdett with her Buff Orpington entry.

Clark Gable and Carol Lombard with their chickens

Famous Ducks

A special tourist attraction for guests of the Peabody Hotel, Memphis, Tennessee, are the five Mallard ducks that enjoy using the unique fountain in the center of the lobby. It is carved from one solid piece of spectacular Italian Travertine marble. At 11:00 a.m. every morning, the ducks ride down the elevator from their enclosed roof-top home called *The Duck Palace* and parade across a red carpet to their private pool. This spectacle is accompanied by the pomp and circumstance these famous ducks so richly deserve. Their journey is made to the accompaniment of a rousing John Philip Sousa march.

Famous ducks in the lobby of the Peabody Hotel

Through the day they entertain the public. At 5 p.m., the ducks cross the red carpet, board the elevator and return to their private abode at the top of the hotel. Without a doubt, these are the most photographed ducks in the world. This ritual dates back to 1930, making Mallard ducks a tradition in this popular hotel for a very long time.

The Peabody Hotel has now included the same performance at their hotel in Orlando, Florida...using different ducks, of course!

Trained Chickens

Can you imagine training a chicken to roller skate? Most of us feel good if we can direct a flock of chickens through a door to close them in at night. Carolyn Steve, Long Island, New York, has been entertaining with her *actor* chickens and rabbits for several years and does 20 shows a year. It takes her about three months to train a chicken using grapes and cheese as

Carolyn Stebe and her trained entertaining chickens on the Regis and Cathy Lee Stage

reward treats.

Carolyn has appeared on television with her chickens and always provides suitable, colorful props that enhance her show. She charges $150.00 per show but reduces the charge for senior citizen groups.

Polish and Cochin Bantams are the easiest to train but Carolyn has trained Mille Fleurs and Antwerp Belgium Quail Bantams as well. She got her idea for training chickens after seeing a parrot show in Florida, where a bird performed on roller skates. Her husband, Roy, makes the special skates for her chickens.

Poultry Photography

In the 1930's, I took many of the chicken pictures that were used in the Stromberg catalog. Because of my lack of experience, I found I was averaging a dozen shots to get one good picture. I hired a professional photographer and by working with him, my ratio improved. One good picture out of six was typical and if I was lucky, one out of three.

At a large show in Columbus, Ohio, I met Bill and Meldra Harris, who are skilled at this specialized photography. Meldra works as Bill's assistant. She can calm a chicken down by talking to it and even goes through motions of rocking it to sleep. Bill has his camera on a tripod. On a table is a measured spot to pose the bird with a suitable curtain for a backdrop. Posing a bird is not easy because the chicken is nervous and patience is required. To get a good picture, Bill and Meldra average 30 minutes for each bird to be photographed.

Many poultry pictures are taken with little thought given to a pose complimentary to the bird. Chickens held by a proud exhibitor are fine, but rarely show the fine quality of the winner. When the owner of the bird wishes to be included in the picture, the bird should be placed on a table of suitable height. If he wishes to hold the bird, it should be standing on his arm or hand. This makes it possible for a complimentary view of the bird and to show good type. Never try to take a picture with others present or in the presence of other chickens, both of which are distractions.

Gary Hart, Australia, learned of my need for pictures for this book and sent me a group which were excellent in quality. He is a poultry judge and loves to take pictures of the winners. Gary sent over 100 beautiful pictures, all complimentary to the breed he photographed. I asked him about his success in taking such wonderful pictures and was told again, patience! It takes time, averaging between 15 to 20 minutes for each picture.

While attending the Hanover Poultry Show in November, 1989, an avid reader of Poultry Press, Frank Glorius, volunteered to be my guide. This doubled the value of seeing a very popular show. At the conclusion, Frank expressed disappointment at some of the pictures in Poultry Press. Too many winners were held by their owners in such bad positions that their true quality and good type did not show. Frank introduced me to Rudi Proll who had a commercial photography booth at the show for those who wanted a picture of their winning bird.

Rudi is a popular photographer because of the quality of his pictures. His photo booth was small, only about six feet wide by twelve feet long. Inside, Rudi had a small table about the size of a card table, upon which was an enclosed lighted stage. The room was dark with only the lights on the stage used to pose the chicken. Chickens waiting to be photographed were kept in cages outside.

There were two cameras on tripods, one for color, one for black and white pictures. These were located about three feet away from the table. With a pointer, Rudi would maneuver the chicken into position, talking to the bird. In a few minutes, it was over and Rudi had made it look very easy. He claimed a 90% success rate and his customers always had a good shot.

You might feel you don't need to build a stage and unless you are taking a lot of pictures, it might not be necessary, but it would make it easier. With little expense, you'll find you have better control over your subject. Remember to take your pictures away from the noise and distractions that abound at a poultry show.

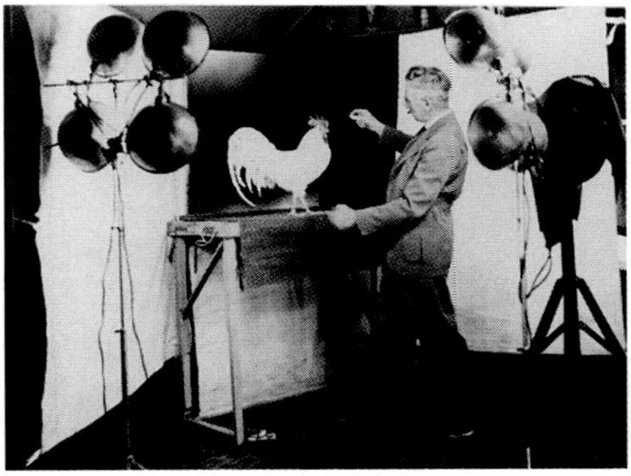

Famous Artist Photographer Louis Schilling at work posing a prize winner

Arthur Shilling, a famous poultry artist, was also a great photographer. He used a very elaborate arrangement of lighting. Arthur was also a touch-up genius when it came to improving pictures. I visited Mrs. Schilling to buy pictures from her and while sorting through hundreds of photographs, I saw a vast number that Arthur had *corrected.* In one, he had painted out the comb, and re-painted in one with five perfect spikes. It was so skillfully done that the final picture didn't show it. In another photo, Arthur painted in a new leg in the proper position. It was impossible to detect this correction in the finished product.

The fancier's magazines published in England and Germany are to be complimented for their selection of poultry pictures which are very impressive. They have nice type winners, but they aren't any nicer than the chickens in the US. It's the extra time they have taken to pose their birds that makes the difference.

It is important that you do your picture taking without an audience and photograph one bird at a time. It's only logical that with more birds in the picture, your complications increase accordingly. If you are photographing a pair or a trio, expect to spend more time and definitely more patience!

When you do have a winner, it's worth the time to photograph it in a manner that makes it look like a winner. Many a great bird has been photographed and looks like a fourth place loser.

Chapter Nine

Folklore

Opposite
'Poultry Facts and Oddities'
by
Hering, Borne, Lamb, Mann
ahd Schroder
Presented courtesy
of
Poultry Tribune
magazine
1930 through 1950

Unusual Poultry Tales

According to some scholars, Mother Goose was a real person. Her name was Elizabeth Vergoose and it is believed that she collected and wrote some of the most famous English nursery rhymes. Her son-in-law, a printer by the name of Thomas Fleet, is credited with publishing the first book of her rhymes in Boston in 1719. Unfortunately, no copies of this book exist to prove the case.

A gosling named Andy was hatched without feet. Many people would have killed it but not Gene Fleming, Hastings, Nebraska. He had compassion for the poor creature and provided it with shoes which in the beginning were hand-crafted. As an adult goose, regular baby shoes worked nicely. Andy got about very naturally and even enjoyed an occasional swim. This unusual story gained worldwide publicity.

Speech was first recorded by brushes, sharp pieces of metal or stone. Then man discovered quills of certain bird feathers made excellent writing tools when their ends were split and sharpened. The Latin word for quill is *penna* or *penne,* thus the word pen. The American Declaration of Independence was signed with a goose quill pen. Many manuscripts of early American history and literature, were written using a goose quill pen. Steel pens were invented in 1820 and replaced feathers.

The goose is regarded as the *feast fowl* in Europe, and is favored over turkey. It's said to be one of the oldest domesticated fowl and was depicted in drawings on the walls of King Tut's tomb. In early days, geese were commonly used as sentries to protect homes against intruders. A gaggle of geese is credited with awaking the Romans to the invading Gauls in 309 B.C.. A distillery in Glasgow, Scotland uses *Geese Police* to guard their inventory of Scotch whiskey.

In China, a favorite parental gift to a bride and groom was a pair of geese. It symbolized a long, faithful marriage. In Europe, A.D. 700, the mother of the bride gave her daughter a goose, not only to perpetuate a faithful marriage but as a sign of guidance for her children. The goose later became known as the *wedding feast bird.*

THE OLDEST KNOWN PICTURE OF A CHICK WAS USED ABOUT 3400 YEARS AGO BY AN ANCIENT EGYPTIAN FISH STORE TO ADVERTISE FISH.

★INSTEAD OF HANDING OUT CIGARS, THE CHINESE PASS EGGS TO THEIR FRIENDS WHEN THEIR WIVES PRESENT THEM WITH MALE CHILDREN.

★A NORTH CAROLINA CHICKEN THIEF STOLE 7 HENS, BUT LOST $150.00 IN CASH WHICH WAS FOUND BY THE POULTRYMAN.

★THE JAPANESE NATIVES OF THE ISLAND OF KOTOSHO WILL NOT EAT EGGS, BELIEVING THEM FATAL TO MAN.

FOR YEARS THE BREAKFAST OF M. BOMMARITO OF DETROIT CONSISTED OF RAW EGGS, LOTS OF THEM. HIS WITNESSED RECORD IS 61 EGGS AT ONE TIME.

★ A HILTON, N.Y. POULTRYMAN MIXES SECRET CHEMICAL COMPOUNDS WITH HIS POULTRY FEED TO PRODUCE COLORED EGG YOLKS.

IN DEERING, MO., TO SETTLE A BET, V. DAVIS KILLED A CHICKEN, COUNTED 8537 FEATHERS.

★IT IS POSSIBLE NOW TO BUY SPECTACLES WITH COLORED LENSES TO FIT BABY CHICKS. THESE GLASSES, DISGUISING ANY BLOODY SCRATCHES ON OTHER CHICKS, PREVENT CANNIBALISM AND FIGHTING.

In Pennsylvania, there is Michaelmas Day, which is a feast day with a bountiful goose dinner. This tradition stems from the religious holiday appointed by Pope Felix III in A.D. 480. In the 15th century, the Feast of St. Michael became popular in England and is still observed by serving a dinner of roast goose. The legend says if you eat a goose on Michaelmas Day, you'll not want for money all year.

In Europe, geese were bred for their large livers. Force-feeding was traditional to produce the extra large livers, which are regarded as a gourmet food and command a very high price in Europe. Specially bred geese for large livers are still produced in Hungary.

Goose fat had many uses. The early Europeans used it in place of butter and it was popular in cooking. Human afflictions also benefited by goose fat. These varied from the treatment of chest colds, winter itch and chapped hands. Some early cosmetics utilized goose fat. American pioneers used it to clean and polish their boots.

Martin is the Patron Saint of geese. This modest man felt unworthy to be made Bishop of Tours and hid among a flock of geese whose noise gave him away. Once discovered, he was crowned Bishop and geese became the traditional food eaten at Martinmas.

The phrase, *goose hangs high*, is said to express the idea everything is fine and probably comes from the fact that wild geese honk as they fly. The height at which they travel is determined by the weather. In storms, the geese fly close to the Earth. In good weather they travel and honk high.

The term *goose step* was originally applied to a British sitting-up exercise. The recruit had to stand on one leg and swing the other backward and forward without bending the knee, just as a goose stands on one leg and swings the other.

In the *Poultry Tribune*, April 1952, I found a news item that said 500 geese were being offered to weed strawberries in the area of Benton Harbor, Michigan. The article states the rental was $2.50 per goose for a period of eight to ten weeks.

For a number of years, the Heart of Missouri Poultry

★IN OCCUPIED FRANCE
URING THE WAR, IF A HEN LAID
EWER THAN 50 EGGS A YEAR
ER OWNER WAS ARRESTED.
THE BA-KALAHARIS OF THE AFRICAN DESERT STORE ALL WATER IN EGG SHELLS, WHICH ARE STORED IN LARGE NUMBERS IN THE SAND. WOMEN SUCK THE WATER FROM THE DEPTH OF THE DESERT SAND THROUGH ONE REED AND, MOUTH FILLED, SQUIRT IT THROUGH ANOTHER ONE INTO THE EGG SHELL.
THERE'S LIFE IN THE OLD FLOCK YET
A CALIFORNIA FLOCK IS MADE UP OF HENS BETWEEN 9 AND 11 YEARS OLD, AND THE HEAD OF THE FLOCK IS A ROOSTER OVER 11 YEARS OF AGE. EGG PRODUCTION EXCEEDS THE AVERAGE OF YOUNGER FLOCKS.
WHICH? ······ OR BOTH ?
AFTER LAYING WELL FOR THREE YEARS, AND BECOMING THE MOTHER OF CHICKS, A HEN BEGAN TO CROW AT 3½ YEARS ··· TOOK ON ALL APPEARANCES OF A MALE, AND AT 4½ YEARS WAS MATED TO A PULLET AND BECAME THE FATHER OF TWO CHICKS. (PROF. W. ROWAN - UNIVERSITY OF ALBERTA)
★IN BALI MEN ARE FORBIDDEN TO OWN OR TEND CHICKENS, THAT PRIVILEGE BEING TRADITIONALLY RESERVED TO WOMEN.
DURING THE CALIFORNIA GOLD RUSH EGGS SOLD FOR AS MUCH AS $50 A DOZEN!
S-H-H-H-H-H
Happy Birthday Henrietta
-A 28-YEAR-OLD OHIO HEN STILL LAYS EGGS.
A POULTRYMAN SHOULD FIGURE TO RAISE ENOUGH PULLETS TO REPLACE 60 TO 70% OF HIS FLOCK. EXPERIMENTS SHOW THAT FOR EVERY PULLET DESIRED IT IS NECESSARY TO START EITHER 5 SETTING EGGS OR 3 BABY CHICKS.
★A CALIFORNIA MAN HAS INVENTED A DEVICE THAT KEEPS ROOSTERS FROM CROWING, BY MAKING IT IMPOSSIBLE FOR THEM TO STRETCH THEIR NECKS.

THE SUCCESS OF CYRUS KING OF PERSIA, IN WELDING TOGETHER HIS EMPIRE ,WAS LARGELY DUE TO HIS MARVELLOUSLY ORGANIZED PIGEON POST, RULING THE GREATER PART OF THE OLD WORLD MORE THAN 500 YEARS BEFORE CHRIST. CYRUS' MESSAGES WERE CARRIED TO AND FROM EVERY CORNER OF HIS EMPIRE BY RACING PIGEONS - WITH A SPEED AND RELIABILITY RIVALING OUR MODERN AIR MAIL SERVICE.
AN OLD REMEDY FOR SORE-EYES: IF YOUR POULTRY HAVE SORE EYES - YOU SHALL TAKE A LEAF OF GROUND-IVY, AND CHAWING IT WELL IN YOUR MOUTH, SUCK OUT THE JUICE AND SPIT IT INTO THE SORE EYE, AND IT WILL MOST ASSUREDLY HEAL IT, AS IT HATH BEEN OFTEN TRIED.
(MARKHAM'S HUSBANDRY, PUBL. A.D.1614)
ARISTOTLE TAUGHT (350 BC) THAT "LONG AND POINTED EGGS ARE FEMALE, THOSE ROUNDED AT THE NARROW END ARE MALE" AND MANY PEOPLE STILL BELIEVE THIS, THO OFTEN PROVED UNTRUE BY EXPERIMENT.
O FEMALE
O MALE
THE DYAKS OF BORNEO PRESENT DISTINGUISHED VISITORS WITH AN EGG - A SORT OF "KEY TO THE CITY"
ILLEGAL FOR ANYONE BUT NOBILITY AND GENTRY TO EAT BAKED CHICKEN
IT WAS ILLEGAL TO EAT BAKED CHICKEN IN SCOTLAND, DURING THE XVI CENTURY, FOR ANYONE BUT NOBILITY AND GENTRY.
JACK TURNER OF HOUSTON, TEXAS PAID $1000 FOR A CAPON, PROBABLY THE WORLD RECORD PRICE FOR A MARKET CHICKEN
$1000
SOME POULTRY FARMS IN FRANCE ARE "MANAGED" EXCLUSIVELY BY THE USE OF A "DIVINING ROD". NO STEP IS TAKEN WITHOUT CONSULTING THE "ROD". IT DETERMINES WHICH BIRDS TO CULL, WHICH TO SELECT FOR BREEDERS, WHICH EGGS OUGHT TO GO INTO THE INCUBATOR, WHICH BIRDS ARE HEALTHY; IT "SELECTS" EVEN THE EQUIPMENT AND FIXTURES.
PROBABLY THE MOST EXPENSIVE WATER FOUNTAINS IN THE WORLD WERE TWO PRICELESS DISHES OF AZTEC ORIGIN USED BY A MEXICAN WOOD CHOPPER TO WATER HIS CHICKENS. AN ARCHEOLOGIST DISCOVERED THESE VESSELS AND BOUGHT THEM FOR 25 CENTS. THEY ARE NOW RECOGNIZED AS SOME OF THE FINEST EXHIBITS OF AZTEC ART.

Farm, Columbia, Missouri and Stahmann Farms, Las Cruces, New Mexico, supplied many goslings as weeder geese. When chemical weed killers were introduced, weeder geese were no longer used.

Superstitions

Predicting the future was done by an East Indian soothsayer employing a dieg shat pylleng or egg breaking board. Using this, he told his clients fortune. After invoking the spirits, he would smash the egg against the board, making the large portion of the shell fall in the middle and the smaller fragments, the good or evil omens, falling to the right or left. If four fragments stayed on the board to form a square, the client was at the point of death.

Ancient Greeks and Romans believed that roosters were so fierce they could frighten lions. So strong was this fear, that a man smeared with broth made from a rooster would not be harmed by lions or panthers.

In Sicily, when caponizing cockerels, the housewife saved the crests she cut off and put them to dry in the sun and wind. The dried crests had the same power as wolf's teeth which the herders saved. When passed over the gums of teething babies, the crests hastened tooth growth without pain or complications.

The Mia-Tze tribe in Asia would not bury a man until the ground was first tested with an egg. While the body was prepared for burial, the male members of the tribe would carry a basket of eggs, dropping them on possible burial sites. If the egg broke when dropped, they moved on until they found a site where the egg did not break. This was a good omen for a suitable burial plot.

A girl who wanted to dream of her future husband would put the first laid egg of a pullet under her pillow.

Rooster testicles were recommended by Alexander of Trolles in the sixth century, to give strength to weak patients.

Cock's testicles were recommended for epilepsy.

Aldrovandi stated that rooster testicles improved lust.

Zoraster thought the rooster was a holy bird that proclaimed early morning and was a symbol of light and sun.

According to Pliny, an early writer, a Roman Princess yearned for a son. To divine whether she would produce a boy, she warmed an egg with her breasts and when she could not tend it, entrusted it to a wet nurse. Between them, they hatched a rooster. The Princess delivered a boy. Had she hatched a pullet, her child would have been a girl.

It was thought that bad habits would depart from a man who drank a solution containing the dried comb of a cock. The smoke of a dried comb of a white or red cock was of benefit to a madman.

Loss of memory was cured by taking a mixture of the gall of a cock and mutton broth. Pain was relieved immediately by taking pills made of pounded, dried flesh of a cock mixed with equal quantities of gallnuts and sumac. Virility was supposed to be greatly increased by an application of cocks' blood mixed with honey.

The cock was not always regarded as a beneficial bird however, demonstrated by the belief that blood drawn in a cockfight, mixed with food and given to a number of people, would cause dissension among them.

There was a superstition that called for placing eggs by doorways as protection to keep the devil out and bring fertility or money into the house.

Mother Carey's chickens refers to a nautical saying about the Albatross which appeared near a ship to warn sailors of an impending storm.

Geese should be plucked in the new moon for better feathers.

Eggs laid on Good Friday should never be eaten. If set under a hen, they were supposed to be superior chick and all cocks.

To hatch male chicks, set eggs under the hen on Tuesday, Thursday, Saturday and Sunday in the forenoon.

To hatch female chicks, set your hen on famine days, Monday and Friday.

It is unlucky to bring eggs into the house after dark.

If fowl roll in the sand, rain is at hand.

IN VENICE, 1556 A.D., THE LUXURY OF SERVING TURKEY WAS PROHIBITED BY LAW
IN TURKEY THE TURKEY IS CALLED THE AMERICAN BIRD
EARLY MEXICANS REVERED THE TURKEY SO HIGHLY, THEY ERECTED MONUMENTS WITH FULL SIZED, GOLD STATUES OF THE BIRD
THE NOTED BIOLOGIST SPALLANZANI (1785 A.D.) STUDIED TURKEY NUTRITION BY AN INGENIOUS METHOD. HE PUSHED A SPONGE, FASTENED TO A STRING, INTO THE DIGESTIVE TRACT. THE DIGESTIVE JUICES SO OBTAINED WERE MIXED WITH VARIOUS FEEDS IN MINUTE CONTAINERS WHICH HE CARRIED IN HIS ARMPITS TO OBSERVE DIGESTIVE PROCESSES AT BODY TEMPERATURE.
PUEBLO INDIANS FORMERLY RAISED TURKEYS ONLY FOR THEIR FEATHERS WHICH THEY USED FOR WAR BONNETS.
AH, TO THINK THAT I SHOULD COME TO SUCH AN END
NEWS
PACKERS MAKING "TURKEY HOT DOGS"
PACKER'S ARE NOW MAKING "TURKEY HOT DOGS". EVIDENTLY THE FIRST APPEARANCE OF THAT HASH IN FOUNDATION GARMENT DESIGNED TO UPLIFT THE TURKEY BUSINESS
A LIVE TURKEY GRACES THE THANKSGIVING TABLE OF THE MICHIGAN VEGETARIAN SOCIETY - AFTER THE CELEBRATION THE THANKFUL NOBLE BIRD JOINS THE DETROIT ZOO
IN THE DAYS OF THE COLONISTS, WILD TURKEYS WERE SO NUMEROUS THEY SOLD FOR 6 CENTS EACH.
BORNE
A TURKEY WITH TWO NECKS IS REPORTED AT SCOBEY, MONTANA
IN ECUADOR TURKEYS ARE "PREPARED" FOR NEW YEAR CELEBRATION BY A STRANGE RITUAL. MADE DRUNK WITH RUM, THEY ARE TIED TO A CORD AND DRIVEN AROUND IN CIRCLES FOR THE AMUSEMENT OF INVITED SPECTATORS
TURKEYS WHICH HAVE HAD THEIR APPENDIX REMOVED ARE NOT SUSCEPTIBLE TO THE DREADED DISEASE KNOWN AS BLACKHEAD.
"THE TERRIBLE TURK" HAD NOTHING ON ME
A GOBBLER ATTACKED AND KILLED A HORSE NEAR DIALLA, EAST PRUSSIA. DURING THE BATTLE THE HORSE WAS SO SHOCKED AND EXCITED THAT IT DIED OF APOPLEXY.

If a cock crows at the front door, someone unexpected is sure to come.

To dream of a hen and chicks is a sure sign you will be married to a widower with many children.

To dream of swimming geese or ducks is a sign you will take a sea voyage.

If a hen lays in a duck's nest, or visa versa, it is most unlucky. Evil will fall on the brood.

Eggs will not hatch if they are set when the wind is blowing from the north.

If a thunderstorm blows in during incubation, the noise created by the thunder will kill every chick.

Chicks that are allowed free range when there is a lot of dew on the ground will get diarrhea and die.

If two hens fight, two female visitors will soon arrive.

It takes less time for chicks to hatch in an incubator than to be hatched under a cluck.

Chicken soup cures the flu.

A hen should be set in the fulling of the Moon if good luck is to be expected.

People in olden times believed red roosters set fires and white roosters could crow mice out of the house.

In England, there was a belief that a crowing rooster could drive away mad ghosts.

In Bohemia, it was believed that if you threw the first and last egg laid by a black hen over the house, the house could not be bewitched.

In Thuringen and Silezie, girls who would like to marry, knock three times against the wall of the chicken house on Christmas and New Year's day, speaking a rhyme. If the rooster crowed they would become a bride the next year. If a hen made a noise, they would wait a long time.

Swallow a gizzard whole and you will become handsome.

Morning egg gathering makes hens lay more than when gathered later in the day.

Eggs set on Sunday produce more roosters.

Eggs set before noon produce more pullets.

It's bad luck to give a chicken away. Something must be paid for it.

IN ABYSSINIA EGG SHELLS ARE USED EXTENSIVELY TO DECORATE THE INTERIOR OF CHURCHES AND THE CROSSES SURMOUNTING THEM.

THE TOWN OF BENDORF (GERMANY) PAYS ANNUAL HOMAGE TO THE EGG BY CELEBRATING AN EGG CARNIVAL. THE CLIMAX OF THE FESTIVAL IS THE CROWNING OF A TALL PINE WITH A CROWN MADE OUT OF TENS OF THOUSANDS OF EGG SHELLS.

(RIGHT) IT'S ONLY BUSTER, A LIVELY RHODE ISLAND RED, OWNED BY WM. LEHR, A RETIRED CIRCUS CLOWN, LOS ANGELES, CAL. BUSTER USES 5-INCH LONG BABY SKATES OF ALUMINUM, TAPED TO HIS BIG FEET!

AT A PHILADELPHIA BALL, MR. AND MRS. SIDNEY HIRST WORE LIVE CHICKENS AS HEADGEAR

EGGS ARE USED TO MEASURE THE CONCENTRATION OF BRINE IN THE MANUFACTURE OF SALT IN CHINA. THE ASH FROM SALT-PLANTS IS BOILED IN A KETTLE WITH SEA WATER. THE LIQUID IS EVAPORATED UNTIL AN EGG FLOATS.

MORE THAN ONE HALF OF THE WHITE IS PUT INTO THE EGG AFTER THE SHELL HAS BEEN FORMED.

FOUR BABY SKUNKS WANDERED INTO A CHICKEN HOUSE, LOOKING FOR COMPANIONSHIP —AND FOUND IT! A HEN STARTED MOTHERING THE SKUNKS— SITTING ON THEM NIGHTLY.

UNDER HENRY VIII EACH LADY OF THE COURT RECEIVED THREE DOZEN EGGS DAILY, ALONG WITH THREE CHICKENS AND EIGHT GALLONS OF BEER. THIS DIET WAS SUPPOSED TO KEEP THEM PLEASINGLY PLUMP!

AN EGG THE MIGHTY VESUVIUS COULD NOT BREAK WHEN POMPEII AND HERCULANEUM WERE DESTROYED, 1900 YEARS AGO, IS DISPLAYED IN A NAPLES MUSEUM.

EGGS BROUGHT $3.00 EACH IN CALIFORNIA GOLD RUSH DAYS

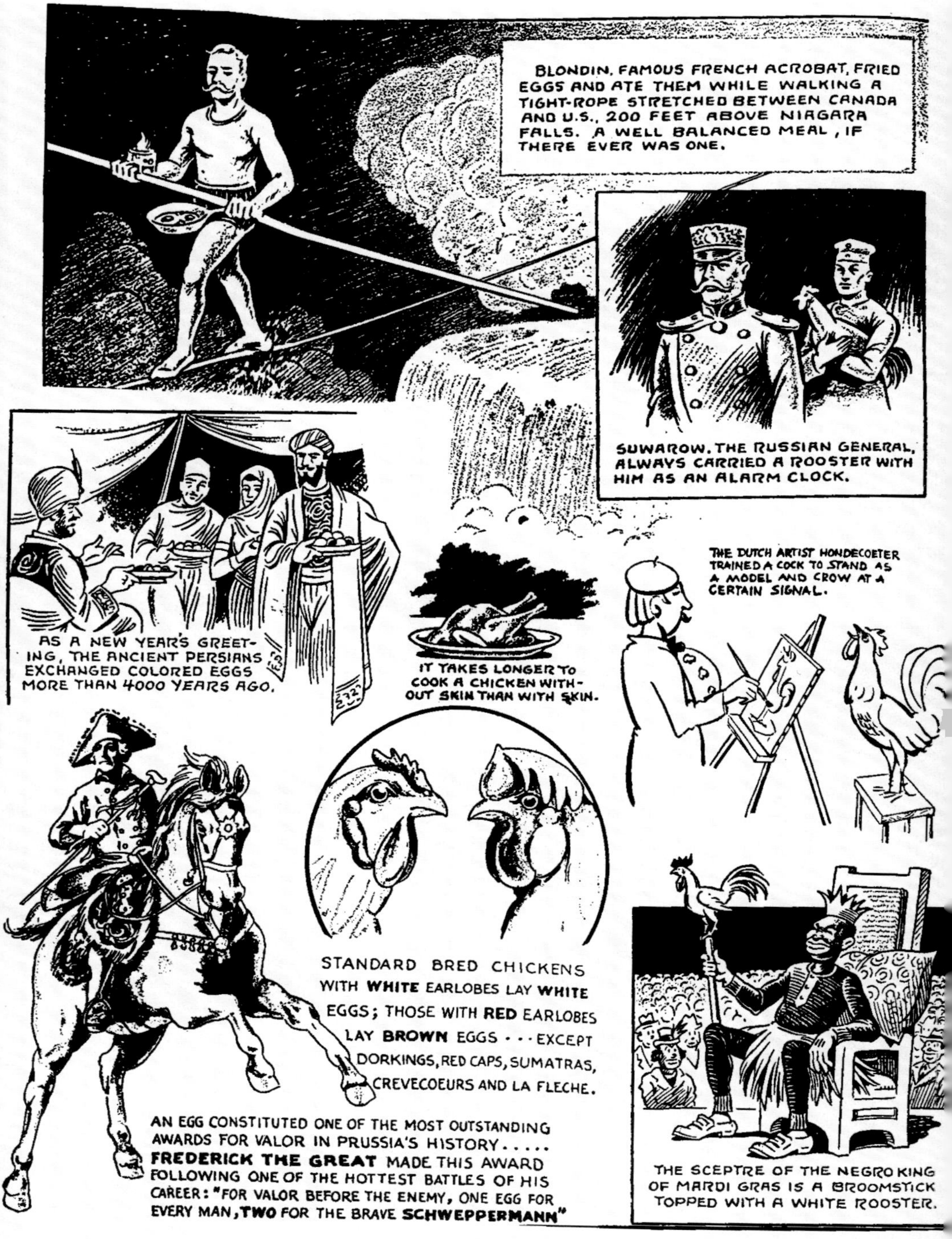
BLONDIN, FAMOUS FRENCH ACROBAT, FRIED EGGS AND ATE THEM WHILE WALKING A TIGHT-ROPE STRETCHED BETWEEN CANADA AND U.S., 200 FEET ABOVE NIAGARA FALLS. A WELL BALANCED MEAL, IF THERE EVER WAS ONE.
SUWAROW, THE RUSSIAN GENERAL, ALWAYS CARRIED A ROOSTER WITH HIM AS AN ALARM CLOCK.
AS A NEW YEAR'S GREETING, THE ANCIENT PERSIANS EXCHANGED COLORED EGGS MORE THAN 4000 YEARS AGO.
IT TAKES LONGER TO COOK A CHICKEN WITHOUT SKIN THAN WITH SKIN.
THE DUTCH ARTIST HONDECOETER TRAINED A COCK TO STAND AS A MODEL AND CROW AT A CERTAIN SIGNAL.
STANDARD BRED CHICKENS WITH WHITE EARLOBES LAY WHITE EGGS; THOSE WITH RED EARLOBES LAY BROWN EGGS . . . EXCEPT DORKINGS, RED CAPS, SUMATRAS, CREVECOEURS AND LA FLECHE.
AN EGG CONSTITUTED ONE OF THE MOST OUTSTANDING AWARDS FOR VALOR IN PRUSSIA'S HISTORY . . . . . FREDERICK THE GREAT MADE THIS AWARD FOLLOWING ONE OF THE HOTTEST BATTLES OF HIS CAREER: "FOR VALOR BEFORE THE ENEMY, ONE EGG FOR EVERY MAN, TWO FOR THE BRAVE SCHWEPPERMANN"
THE SCEPTRE OF THE NEGRO KING OF MARDI GRAS IS A BROOMSTICK TOPPED WITH A WHITE ROOSTER.

A horseshoe hung in the chimney will protect chickens from hawks.

It is a sign of rain if hens spread and ruffle their tail feathers.

If a rooster crows at night there will be rain before morning.

If birds or ducks nest on high ground, expect a wet summer. Expect a dry season if the nest is close to the edge of a stream.

Poultry Sayings

Quaker ladies would not curse. When angry they would say, *Cussed Minorcas!*

He talks like a chicken without a head.

Better to have a half an egg than an empty shell.

No rooster is crowing to that. *(Nobody shall know that.)*

Be there like chickens. *(Fast.)*

To write cocklegs. *(Very bad.)*

To roost with the chickens. *(Go to bed early.)*

As dizzy as a goose. *(Silly as a goose.)*

I have a goose to pluck with you. *(A quarrel.)*

To pluck a goose without making it cry. *(To rob, little by little.)*

Geese can be plucked as long as they have feathers. *(To cheat people of their goods.)*

Wild goose chase. *(Spend valuable time for nothing.)*

When ducks walk on snow at Thanksgiving they will walk in mud on Christmas.

Children made to stay indoors are said to be cooped up.

The pecking order means the company President yells at the Vice-President, the Vice-President yells at the secretary, etc. This is a modern version of the peck ing order. Or, you can peck at your wife's sister, but you dare not peck at your mother-in-law.

When people meet to plan an illegal or questionable activity, they are hatching a plot.

When the joke falls flat, you are said to have laid an egg. etc.

The tour leader who takes good care of his group might be called a mother hen.

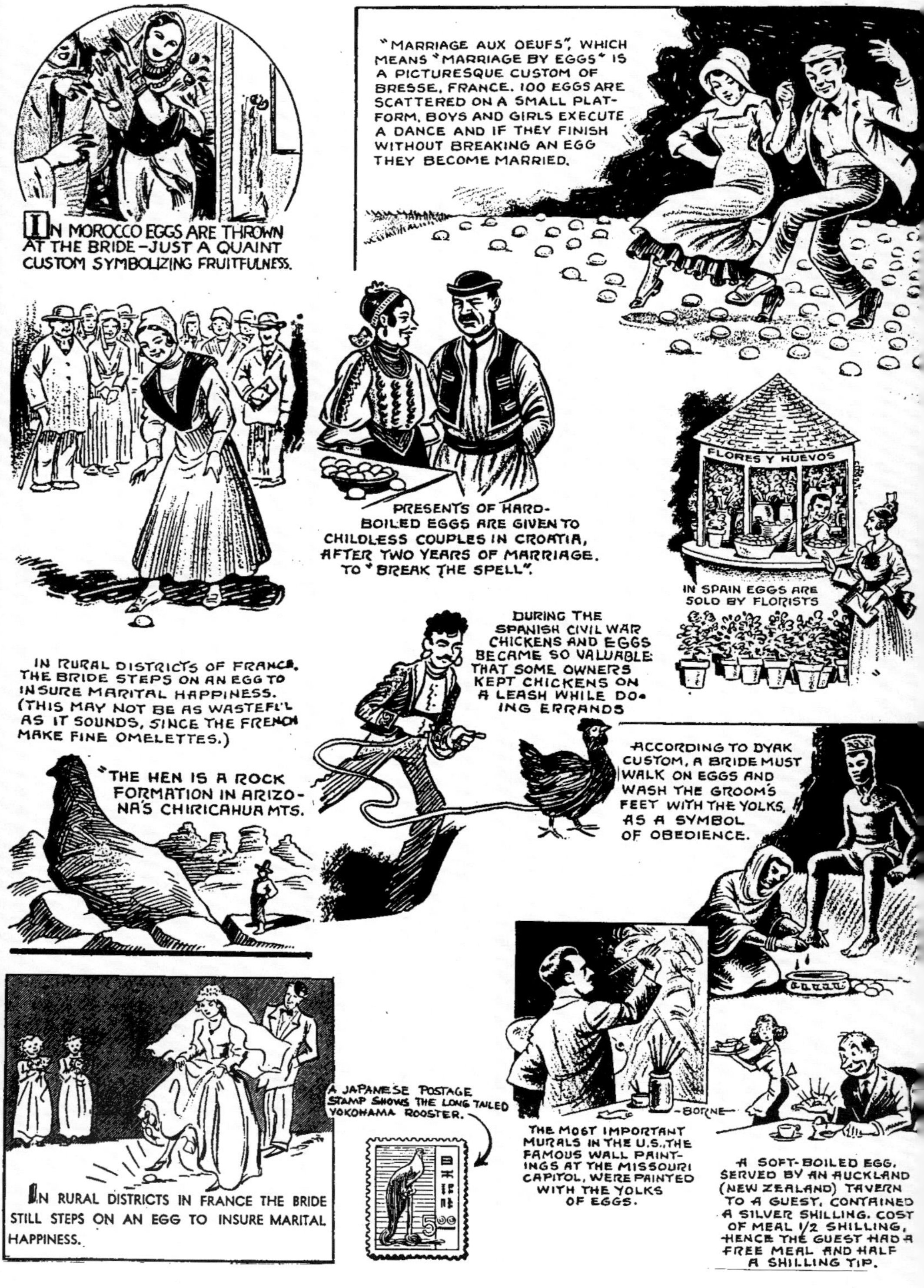
IN MOROCCO EGGS ARE THROWN AT THE BRIDE–JUST A QUAINT CUSTOM SYMBOLIZING FRUITFULNESS.
"MARRIAGE AUX OEUFS", WHICH MEANS "MARRIAGE BY EGGS" IS A PICTURESQUE CUSTOM OF BRESSE, FRANCE. 100 EGGS ARE SCATTERED ON A SMALL PLATFORM, BOYS AND GIRLS EXECUTE A DANCE AND IF THEY FINISH WITHOUT BREAKING AN EGG THEY BECOME MARRIED.
PRESENTS OF HARD-BOILED EGGS ARE GIVEN TO CHILDLESS COUPLES IN CROATIA, AFTER TWO YEARS OF MARRIAGE, TO "BREAK THE SPELL".
FLORES Y HUEVOS
IN SPAIN EGGS ARE SOLD BY FLORISTS
IN RURAL DISTRICTS OF FRANCE, THE BRIDE STEPS ON AN EGG TO INSURE MARITAL HAPPINESS. (THIS MAY NOT BE AS WASTEFUL AS IT SOUNDS, SINCE THE FRENCH MAKE FINE OMELETTES.)
DURING THE SPANISH CIVIL WAR CHICKENS AND EGGS BECAME SO VALUABLE THAT SOME OWNERS KEPT CHICKENS ON A LEASH WHILE DOING ERRANDS
"THE HEN IS A ROCK FORMATION IN ARIZONA'S CHIRICAHUA MTS.
ACCORDING TO DYAK CUSTOM, A BRIDE MUST WALK ON EGGS AND WASH THE GROOM'S FEET WITH THE YOLKS, AS A SYMBOL OF OBEDIENCE.
IN RURAL DISTRICTS IN FRANCE THE BRIDE STILL STEPS ON AN EGG TO INSURE MARITAL HAPPINESS.
A JAPANESE POSTAGE STAMP SHOWS THE LONG TAILED YOKOHAMA ROOSTER.
–BORNE–
THE MOST IMPORTANT MURALS IN THE U.S., THE FAMOUS WALL PAINTINGS AT THE MISSOURI CAPITOL, WERE PAINTED WITH THE YOLKS OF EGGS.
A SOFT-BOILED EGG, SERVED BY AN AUCKLAND (NEW ZEALAND) TAVERN TO A GUEST, CONTAINED A SILVER SHILLING. COST OF MEAL 1/2 SHILLING, HENCE THE GUEST HAD A FREE MEAL AND HALF A SHILLING TIP.

BY CROSSING DUCKS WITH CHICKENS OR GUINEA FOWL, ALBERT COMBE OF OBERDERDINGEN, GERMANY PRODUCED ODD-LOOKING CREATURES. THE CROSSBREDS HAD INCOMPLETE WEB FEET. ILLUSTRATIONS ARE EXACT REPRODUCTIONS (D. LANDW. GEFL. ZTG., VOL. 34, PAGE 611)
A HEN PLAYS AN IMPORTANT PART IN AMERICA'S OUTSTANDING GRAND OPERA!
"IM MARRYING A MAN OF WEALTH! HE OWNS A DOZEN CHICKENS."
★IN A MIDWEST STATE, TWO PERFECTLY FORMED HEALTHY BABY CHICKS HATCHED FROM A SINGLE EGG PLACED AMONG OTHERS IN AN INCUBATOR.
IN SIAM CHICKENS ARE SOMETIMES GROWN INSIDE A BAMBOO STICK, PRODUCING ELONGATED BIRDS SUPPOSEDLY OF ESPECIALLY FINE FLAVOR AND QUALITY.
★IN 1778 CHICKENS WERE SELLING AT $50.00 EACH IN NEW YORK. EGGS WERE $9.00 A DOZEN.
"I'M FOREVER BLOWING BUBBLES!"
IN MOROCCO EGGS ARE THROWN AT THE BRIDE—JUST A QUAINT CUSTOM SYMBOLIZING FRUITFULNESS.
—BORNE—
MONTANUS, FAMOUS GERMAN MONK OF THE MIDDLE AGES, BELIEVED HIMSELF TO BE A GRAIN OF WHEAT. AFRAID OF BEING PICKED UP BY A CHICKEN, HE NEVER LEFT HIS HOUSE.
★THE ZOOLOGICAL DEPT. OF THE UNIVERSITY OF NEBRASKA REPORTS THAT A DEAD DUCK IN HACKBERRY LAKE TURNED TO SOAP AS A RESULT OF THE ACTION OF THE ALKALI WATER UPON THE DUCK'S FAT.
BOMBO, THE TRAINED R.I. RED CHICKEN, HAS LEARNED TO WALK A STRAIGHT LINE FOR LONG DISTANCES.
MANY HOUSEWIVES AT MILAN, ITALY, WHEN PREPARING DRESSED CHICKENS RECENTLY WERE SURPRISED TO FIND THEM STUFFED WITH STOCK CERTIFICATES WORTH LARGE SUMS OF MONEY. AS A RESULT, A GANG OF "STOCK BROKERS" WAS UNCOVERED WHO HAD DONE A THRIVING BUSINESS OF SMUGGLING INTO AND OUT OF ITALY STOCKS AND BONDS HIDDEN IN CHICKENS.

THREE DAYS AFTER
THE BIRTH OF A CHILD, THE CHINESE SEND RED-COLORED EGGS TO RELATIVES-ODD NUMBERS WHEN IT IS A SON, EVEN NUMBERS WHEN IT IS A GIRL.
LAMB
"NOW, I'VE GROWN UP!"
★A BABY CHICK AT A WESTERN EXPERIMENTAL STATION CROWED AT THE AGE OF NINE DAYS!
I HATE TO BE SEEN IN THAT OLD JALOPPY. WISH THEY'D GET A NEW MODEL
IT'S THE MACHINE AGE, ALL RIGHT. A HEN WHICH REFUSES TO LAY ANYWHERE EXCEPT IN AUTOMOBILES IS OWNED BY MRS. S. T. DAVIS, FARMERSVILLE, LA.
MR. J. G. RAND CRIPPLED A WIL GOOSE BY A SHOT IN THE WIN TAKEN HOME, THE GOOSE BECAME TAME. SHE LEARNE TO DRIVE THE COWS HOME FR PASTURE AND INTO THE STALL SHE MADE OBSTINATE COWS OB BY BITES IN THE LEG. AFTER YEAR MATERNAL DUTIES FORCED HER TO RELINQUIS THE JOB AS "COW DOG"
"RIVER, STAY 'WAY FROM MY DOOR"
★EARLY ITALIAN PAINTERS USED EGG YOLK FOR PAINTING GIRLS' FACES. EGGS WITH DARK YOLKS WERE USED FOR COUNTRY GIRLS ..... AND LIGHTER YOLKED EGGS FOR CITY GIRLS.
★DURING A RECENT CALIFORNIA FLOOD, A PLYMOUTH ROCK HEN RODE A LOG 200 MILES DOWN THE FEATHER AND SACRAMENTO RIVERS BEFORE BEING RESCUED-- UNHARMED.
A GOOSE WILL MAKE A GOOD PET, ACCORDING TO A MISSOURI SALESMAN, WHO HAS ONE THAT FOLLOWS HIM AROUND LIKE A DOG. THIS WISE GOOSE RIGIDLY OBSERVES TRAFFIC LAWS AND STOPS FOR RED LIGHTS.
E.M.
M. PASQUETTI INFLATED THE SIZE AND PRICE OF HIS TURKEYS WITH A TIRE PUMP. THE FRAUD WAS DISCOVERED WHEN ONE TURKEY EXPLODED AND THE EYES OF A CUSTOMER WERE INJURED BY FLYING BONES
A LITTLE ROOSTER WAS RECENTLY "CAUGHT BETWEEN THE DEVIL AND THE DEEP BLUE SEA." WITH TWO DOGS AFTER HIM, HE HEADED STRAIGHT FOR THE RIVER, KEPT RIGHT ON GOING, SWAM THE BRINK IN TWO MINUTES FLAT, AND LEFT HIS BEFUDDLED PURSUERS ON THE OPPOSITE BANK.

THE FATHER OF HIS COUNTRY IS ALSO THE FATHER OF THE AMERICAN GOOSE INDUSTRY. GEORGE WASHINGTON IMPORTED THE FIRST GEESE TO THE UNITED STATES.
ROASTED GOOSE, SERVED AND EATEN ALIVE, WAS CONSIDERED A DELICACY DURING THE MIDDLE AGES
To My Valentine
A KANSAS STATE COLLEGE HEN LAID AN EGG CONTAINING A VALENTINE MESSAGE WRITTEN BY M. A. SMITH, KANSAS CITY, FOR HIS WIFE.
(CONTRIBUTED BY MRS. DORA L. THOMPSON, MANHATTAN, KANSAS)
GEESE INFLUENCED THE DESTINIES OF SEVERAL EUROPEAN NATIONS WHEN THEIR NOISE AWAKENED THE GARRISON IN THE CAPITOL OF ANCIENT ROME (390 B.C.) AND FRUSTRATED THE ATTACK BY THE BELEAGUERING GAULS.
TAKE IT FROM WALTER WINCHELL—THERE IS A STEAM HEATED DUCK POND AT ROSLYN, L.I. IT IS THE PRIVATE PROPERTY OF A MILLIONAIRE WHOSE DUCKS CAUGHT COLD.
A RECENT LAW REQUIRES ALL GOVERNMENT EMPLOYEES IN LITHUANIA TO EAT ROAST GOOSE EVERY SUNDAY.
DUCKS AND GEESE ON ONE OF ENGLAND'S LARGEST WATERFOWL FARMS OBSERVE TRAFFIC LIGHTS
IN BABYLON A STONE DUCK WAS THE STANDARD OF WEIGHT.
WHICH CAME FIRST: THE CHICKEN OR THE EGG?
CHICKEN, ACCORDING TO THE GOOD BOOK
GENESIS 1.21
BORNE

YES, HENRY CROWS A LOT BUT IT DOESN'T MEAN ANYTHING
ONE CAN'T BE TOO CAREFUL WHAT ONE SAYS AROUND THOSE CHILDREN
CHICKENS COMMUNICATE WITH EACH OTHER BY DEFINITE SOUND SIGNALS. THE TWO SEXES HAVE TWO DIFFERENT SETS OF SIGNALS, NOT INCLUDING CROWING, WHICH IS NOT CONSIDERED A COMMUNICATIVE SIGNAL
CHICKS RAISED IN COMPLETE ISOLATION FROM OLDER BIRDS LEARN THIS SYSTEM OF COMMUNICATION AT AN EARLIER AGE
"I PREFER ROASTED CHICKEN."
"I'LL TAKE MINE BROILED."
★IN ANCIENT ROME, ONE OF THE COOKING TRICKS WAS TO SERVE A CHICKEN ROASTED ON ONE SIDE AND BROILED ON THE OTHER
YOU GIRLS CAN'T DO A *THING* / CAN'T DO
A VIRGINIA FARM BOY TOLD HIS EMPLOYER THAT A ROOSTER HAD LAID AN EGG. THE FARMER, NOT BELIEVING THE BOY, PENNED THE ROOSTER UP. HE LAID 5 MORE EGGS!
THE PIKE COUNTY POULTRY ASSN. WILL MEET SATURDAY. BRING COCKERELS, CAPONIZING TOOLS AND SANDWICHES FOR LUNCH
MATTER OF TASTE OR DIGESTION?
IN NIGERIA
LIVE CHICKENS ARE MARKETED AFTER ALL FEATHERS ARE PLUCKED SO THE CUSTOMER SEES WHAT HE BUYS
THE ANCIENTS BELIEVED THAT EVERY COCK LAID ONE EGG DURING HIS LIFE, BUT THAT IT WAS VERY SMALL. (PRIZE WINNING ITEM BY A. M. MOVIUS, LEDGERWOOD, N.D.)
WELL, ANYHOW, THEY DON'T GRUMBLE ABOUT MY COOKING
H. RABINOWITZ OF PHILADELPHIA DIVORCED HIS WIFE BECAUSE SHE MADE HIM EAT AT THE SAME TABLE WITH A COUPLE OF PET HENS
NOW-REMEMBER THIS DATE!
GOOD FRIDAY
EGGS LAID ON GOOD FRIDAY KEEP FRESH ALL YEAR AND BRING THEIR OWNER LUCK, ACCORDING TO AN ENGLISH SUPERSTITION.
THERE'S A GOOD YEAR'S WORK.
AN EGG-LAYING RECORD WAS MADE BY A BLACK ORPINGTON IN NEW ZEALAND WHO, IN 1930, LAID 363 EGGS!
A DUCK LAID 18 BLACK EGGS (THE DUCK IS OWNED AND THE ITEM CONTRIBUTED BY MR. G. DODD, R.#1, BRISCOE, TEXAS.
AH, HATCH 'EM YOURSELF, I'M GOING HOME!
A BROODY HEN WAS TAKEN ACROSS A RIVER, 330 YARDS WIDE, TO HATCH A CLUTC OF EGGS. SHE MUST HAVE BEEN "HOMESICK BECAUSE AN HOUR LATER SHE SWAM THE RIVER AND WENT STRAIGHT TO HER NEST.

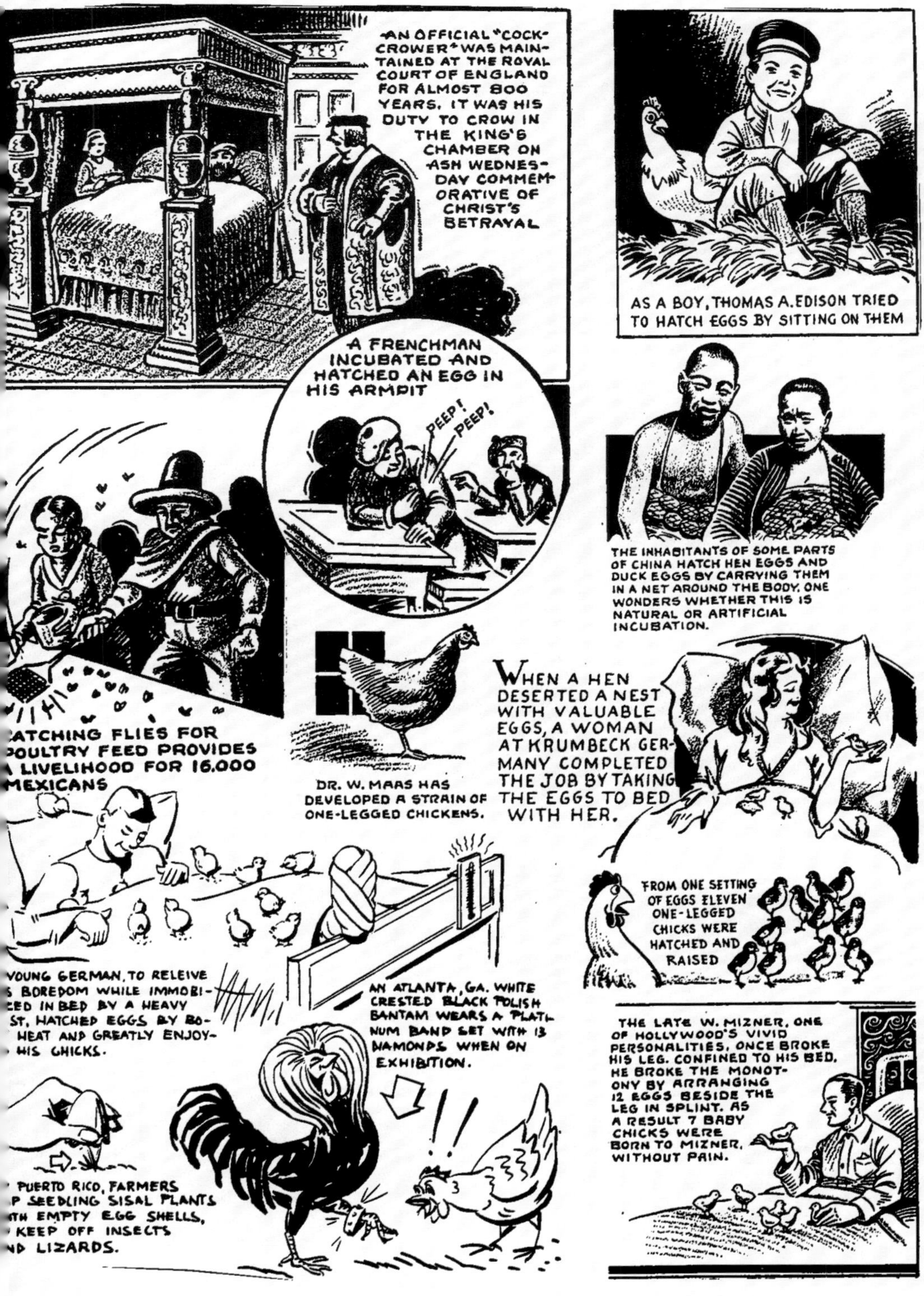
AN OFFICIAL "COCK-CROWER" WAS MAINTAINED AT THE ROYAL COURT OF ENGLAND FOR ALMOST 800 YEARS. IT WAS HIS DUTY TO CROW IN THE KING'S CHAMBER ON ASH WEDNESDAY COMMEMORATIVE OF CHRIST'S BETRAYAL
AS A BOY, THOMAS A. EDISON TRIED TO HATCH EGGS BY SITTING ON THEM
A FRENCHMAN INCUBATED AND HATCHED AN EGG IN HIS ARMPIT
PEEP! PEEP!
THE INHABITANTS OF SOME PARTS OF CHINA HATCH HEN EGGS AND DUCK EGGS BY CARRYING THEM IN A NET AROUND THE BODY. ONE WONDERS WHETHER THIS IS NATURAL OR ARTIFICIAL INCUBATION.
CATCHING FLIES FOR POULTRY FEED PROVIDES A LIVELIHOOD FOR 16,000 MEXICANS
DR. W. MAAS HAS DEVELOPED A STRAIN OF ONE-LEGGED CHICKENS.
WHEN A HEN DESERTED A NEST WITH VALUABLE EGGS, A WOMAN AT KRUMBECK GERMANY COMPLETED THE JOB BY TAKING THE EGGS TO BED WITH HER.
FROM ONE SETTING OF EGGS ELEVEN ONE-LEGGED CHICKS WERE HATCHED AND RAISED
YOUNG GERMAN, TO RELEIVE S BOREDOM WHILE IMMOBI-ZED IN BED BY A HEAVY ST, HATCHED EGGS BY BO- HEAT AND GREATLY ENJOY- HIS CHICKS.
AN ATLANTA, GA. WHITE CRESTED BLACK POLISH BANTAM WEARS A PLATINUM BAND SET WITH 13 DIAMONDS WHEN ON EXHIBITION.
THE LATE W. MIZNER, ONE OF HOLLYWOOD'S VIVID PERSONALITIES, ONCE BROKE HIS LEG. CONFINED TO HIS BED, HE BROKE THE MONOTONY BY ARRANGING 12 EGGS BESIDE THE LEG IN SPLINT. AS A RESULT 7 BABY CHICKS WERE BORN TO MIZNER, WITHOUT PAIN.
PUERTO RICO, FARMERS P SEEDLING SISAL PLANTS ITH EMPTY EGG SHELLS, KEEP OFF INSECTS ND LIZARDS.

THE CACKLING OF A HEN WAS CONSIDERED AN ILL OMEN BY SOME ANCIENTS. THEY ATE HENS SO THERE WOULD BE FEWER OF THEM TO CACKLE
CHICKEN MEAT IS USED IN GERMANY TO MANUFACTURE CARNOFIL, WHICH IS A CLOTH SUBSTITUTE FROM WHICH SUITS ARE MADE.
GEN. ITALO BALBO, WHO LED A FLEET OF AIRPLANES ACROSS THE ATLANTIC
THE ITALIAN GOVERNMENT PROHIBITS CHICKEN AS PROVISION FOR AVIATORS. TO ITALIAN AIRMEN, FOWL IS A JINX, BECAUSE IT CANNOT FLY WELL
THE CHINESE MAINTAIN "EGG CEMETERIES", WHERE THEY BURY THEIR EGGS. TOMBSTONE-LIKE MARKERS IDENTIFY THE OWNER AND GIVE DATE OF BURIAL. WHEN DUG UP A FEW YEARS LATER THESE EGGS ARE CONSIDERED A DELICACY. EGGS 10 YEARS OLD OR OLDER ARE SAVED FOR THE GRANDEST CEREMONIES.
"A MIGHTY OAK FROM A LITTLE CHICKEN GREW": ONE OF THE ROCK HENS BELONGING TO LELAND LAPP, CUNO, TEXAS, SPROUTED AN OAK FROM ITS CROP
REX STOUT, FAMED AUTHOR, FEEDS HIS CHICKENS BLUEBERRIES BELIEVING IT IMPROVES FLAVOR OF THE CHICKEN MEAT
EGG SMELLING IS A HIGHLY PRIZED ART IN CHINESE EGG BREAKING ESTABLISHMENTS
HEDGE DISPLAYS ON DR. ERDMANN'S ESTATE AT EAST HAMPTON ARE SCULPTURED INTO ROOSTERS. "CLUCKS" WITH CHICKS, DUCKS GEESE AND OTHER KINDS OF FOWL.
IN SOME SECTIONS OF THE ORIENT, EGGS ARE TESTED BY HOLDING THREE EGGS IN THE MANNER SHOWN BY THE ILLUSTRATION. IF THE CENTER EGG IS FRESH, IT REVOLVES SLOWLY AROUND ITS AXIS. (CONTRIBUTED BY A.R. WESTERDAHL, LA GLORIA, CUBA.)

COCK FIGHTS WERE STAGED INSIDE THE CHURCH AS A FAVORITE AMUSEMENT AT FESTIVALS AND WAKES IN ENGLAND AND WALES, SOME CENTURIES AGO

A CHICK WHICH SINGS LIKE A CANARY IS OWNED BY MR. WALSH, 7148 ALASKA ST., DETROIT, MICHIGAN.

IN THE MIDDLE AGES COCK FIGHTING WAS A LEGITIMATE AMUSEMENT AT WAKES AND IN SCHOOLS OF WALES. COCKING-MAINS WERE PROMOTED BY THE CHURCH AND TOOK PLACE IN CHURCHES.

ON THE WOODSIDE POULTRY FARM AT NEENAH, WISC., A ROOSTER HAS LEARNED TO SMOKE CIGARETTES.

TAKING ADVANTAGE OF RECENT LOWER FEED PRICES IN HOLLAND, GERMAN FLOCK OWNERS ON THE HOLLANDISH BORDER LET THEIR CHICKENS CROSS THE BORDER TO EAT CHEAP FEED IN HOLLAND, BUT LAY HIGH PRICED EGGS IN GERMANY.

AFTER WHIPPING EVERY ROOSTER IN THE FIELD, THIS GAME COCK CAME HOME, MOUNTED THE WELL CURB, AND SAW HIS REFLECTION IN THE WATER BELOW. HE JUMPED FOR HIS NEXT VICTIM, AND DROWNED BEFORE HE COULD BE FISHED OUT.

WITH EVERY DUCK SERVED AT THE CAFE ANGLAIS, PARIS' OLDEST RESTAURANT- ESTABLISHED IN 1582- THE GUEST IS PRESENTED WITH A PICTURE OF A BEARDED GENTLEMAN AND WITH THE NUMBER OF DUCKS SERVED SINCE 1890.

IT HAS LONG BEEN A CUSTOM IN SOME COUNTRIES TO HAVE COCKS FIGHT BEFORE THEY ARE KILLED TO MAKE THEM MORE TENDER

A NEW BULGARIAN POSTAGE STAMP PUBLICIZES POULTRY

ONE OF THE LARGEST WAGERS IN HISTORY WAS MADE ON THE RESULT OF A COCK-FIGHT, BRANDON, DUKE OF SUFFOLK, BET AND LOST THE INCOME TAX BESTOWED UPON HIM FOR LIFE BY HIS ROYAL BROTHER-IN-LAW, HENRY VIII

IT USED TO BE THE CUSTOM FOLLOWING COCK FIGHTS TO MIX DRINKS WHICH CONTAINED AS MANY INGREDIENTS AS THE VICTORIOUS COCK HAD TAIL-FEATHERS. THIS IS ONE VERSION OF THE ORIGIN OF "COCKTAIL".

IN SOME PARTS OF INDIA FIGHTING COCKS ARE HATCHED UNDER PIGEONS IN THE BELIEF THAT PIGEON-HATCHED STOCK IS MORE PUGNACIOUS AND HARDIER.

A BARRED ROCK HEN OF MRS. G. HUDDLESTON, KITTERY POINT, ME., LAID ONLY EMPTY EGG SHELLS (OFF THE GOLD STANDARD)

HENRY IV, KING OF FRANCE, WANTED EVERYONE TO "HAVE A CHICKEN IN THE POT ON SUNDAY"!
F.H. LAMB
★ A MISSOURI WOMAN HAS A HEN THAT COMES IN AND SITS WITH HER EVERY EVENING WHILE THE WOMAN CROCHETS.
"WHATCHA DOIN'?"
"I'M PUNCTURIN' PUFFY!"
PSST-T-T
★ A MEMPHIS, TENN. FARMER HAS A HEN NAMED PUFFY. EACH WEEK PUFFY SWELLS UP LIKE A BALLOON --THE FARMER PUNCTURES HER WITH A PENKNIFE AND SHE IS NORMAL AGAIN. THIS HAS CONTINUED FOR WEEKS.
IN PERSIA MORE EGGS ARE CONSUMED PER CAPITA THAN ANYWHERE ELSE IN THE WORLD. IN SPORT CLUBS IT IS NOT UNCOMMON TO EAT, AFTER EXERCISES, 40 OR MORE EGGS PER PERSON
CHICKENS CAN SEE HEAT.
FAT 1 OZ.
WATER 1 OZ.
IN DIGESTING AN OUNCE OF FAT A CHICKEN MANUFACTURES MORE THAN AN OUNCE OF WATER IN ITS INTERNAL LABORATORY.
UNFAIR TO ORGANIZED LAYERS
★ AN INDIANA HEN IS LAYING DOUBLE EGGS --ONE COMPLETE EGG, SHELL AND ALL, INSIDE ANOTHER EGG.
A HITCH-HIKING CHICKEN "RODE THE RODS" UNDER A PULLMAN CAR OVER 2000 MILES FROM NEW ORLEANS TO LOS ANGELES.
★ IN MOROCCO EGGS ARE THROWN AT NEWLYWEDS, JUST AS WE THROW RICE. THE EGG IS A SYMBOL OF FRUITFULNESS.
AFTER THE BRITISH BOMBARDED LEWES, DEL., FOR 22 HOURS APRIL 6, 1813, THE CASUALTIES WERE: ONE CHICKEN KILLED, ONE PIG WOUNDED

BY ROYAL GRANT, 300 YEARS AGO, THE SALT WORKERS OF HALLE WERE PERMITTED TO COOK AND STORE EGGS IN THE BRINE OF THE ROYAL SALT WORKS. EVER SINCE, ON NEW YEAR'S DAY, THE RULERS OF GERMANY HAVE RECEIVED AS A TRIBUTE, THREE DOZEN OF THESE SALT EGGS, WHICH ARE VALUED FOR THEIR SPECIAL FLAVOR. THE SALT WORKERS WEAR THE SAME QUAINT STYLE OF DRESS AS THEIR FOREFATHERS WORE CENTURIES AGO.
15 POUNDS OF DRESSING!
MRS. EDNA MAE POTTER ATE TEN FULLY GROWN CHICKENS, THUS WINNING THE LOS ANGELES COUNTY CHICKEN EATING CONTEST.
COMMUNITY TOMS ARE USED IN SOME LOCALITIES OF CENTRAL EUROPE MUCH IN THE MANNER AS COMMUNITY BULLS. THE VILLAGE-OWNED TOMS ARE LEFT WITH THE TURKEY RAISER FOR A FEW DAYS, ONCE OR TWICE DURING THE MATING SEASON.
AT OAK PARK, ILL. IT IS AGAINST THE LAW FOR A ROOSTER TO CROW BEFORE 6 A.M.
THE SWEDISH SPATTA CAKE IS MADE OF 80 EGGS, POTATOES, FLOUR, AND SUGAR. IT TAKES TWO PERSONS TO MAKE IT—ONE TO TURN THE SPINDLE AND ONE TO DROP SMALL DABS OF DOUGH ON THE SPINDLE.
PUEBLO INDIANS FORMERLY RAISED TURKEYS ONLY FOR THEIR FEATHERS WHICH THEY USED FOR WAR BONNETS.
BY LAW THE COLOMBIAN GOVERNMENT OWNS THE GIZZARD OF EVERY BIRD IN THE EMERALD MINING REGIONS. ARMED GUARDS WATCH THE FLOCKS. AT KILLING TIME OFFICIALS EXAMINE THE GIZZARD FOR EMERALDS.
A THIMBLE WAS FOUND IN AN OTHERWISE PERFECT EGG!
AT THE START OF THE RAINY SEASON, THE EAST INDIAN MUNDAS FASTEN A FAT CHICKEN BY ONE WING TO A TALL BAMBOO POLE. WITH GREAT CEREMONY THE POLE IS THEN ERECTED ON A MANURE PILE TO ASSURE FERTILITY OF THE LAND.

A man who thinks deeply is said to be brooding.
Scarce as hen's teeth. *(Chickens don't have teeth but it indicates something is extremely scarce.)*
Sociable as chickens in a coop.
Cross as a setting hen.
Every proper farmer goes to bed with the chickens and gets up with the crowing of the cocks.
Hen house. *(The house controlled by the woman.)*
Hen-pecked.
(Husband is dominated by the woman.)
Cock-sure. *(Very positive.)*
Cock-sure women and hen-sure men.
(Proverb about the war between the sexes.)
Cock's comb.
(A conceited fool, show-off, flashy dresser.)
Hens like to lay where they see an egg.
Running around like a chicken with its head cut off.
To egg someone on is to give them encouragement.
What is sauce for the goose, is sauce for the gander.
(Tit for tat.)
Goose quill gentleman. *(A lawyer with a goose quill pen behind his ear.)*
It is a silly goose that comes to a fox's sermon.
Never trust a kitten to watch the chicks.
One fowl got is better than two promised.
It is of no use to lock up when the fowls are stolen.
The early bird gets the worm.
Gather your eggs at noon, you gather them too soon.
Mayflower chickens are sure to thrive. Blackberry chickens you can't keep alive.
It's a bad hen that lays away from the farm.
A black hen can lay white eggs but a white hen can not lay black eggs.
A servant and a cock should be kept but a year.
He is like a cock, always crowing.
Mad as a wet hen. *(Angry.)*
Swan song. *(A final effort, accomplishment.)*
Talking turkey. *(Getting down to business.)*
Dreams of chickens entering your house.
(You will attain wealth and honor.)
Dreams of cockfighting. *(Signifies wrangles and quarrels.)*

CATTERING THE ASHES OF THE
ULE LOG IN THE CHICKEN
OUSE KEEPS AWAY
ISEASE ACCORDING TO
N ENGLISH CHRISTMAS
UPERSTITION.
SEVEN PAIRS OF TURKEYS SURVIVED THE THE FLOOD IN NOAH'S ARK, NOT ONE PAIR AS IS POPULARLY BELIEVED. GEN.7.4
"YOU'LL HEAR FROM THE UNION ABOUT THIS!"
★A WHITE LEGHORN HEN IN THE SOUTHWEST LAID SEVEN EGGS IN THREE HOURS.
A NEBRASKA FARMER CHOPPED OFF A CHICKEN'S HEAD, BUT THE CHICKEN JUMPED UP AND RAN AWAY, RETURNING TWO DAYS LATER, WHEN THE FARMER KILLED IT FOR KEEPS.
A ROOSTER PERCHED OUTSIDE THE BUILDING IS THE TRADITIONAL DISPLAY SIGN OF POULTRY DEALERS IN PORTUGAL
★ AN ENGLISHMAN'S HOBBY IS PAINTING PORTRAITS ON EMPTY EGG SHELLS. HE HAS PAINTED SOME OF THE WORLD'S MOST FAMOUS PEOPLE IN THIS WAY.
A DOMESTIC BREED OF DUCKS, POPULAR IN CENTRAL GERMANY BUILDS ITS NESTS ON POLES AND TREES.
THE PHOENIX
A DOMESTIC CHICKEN OF JAPAN, DEVELOPS A LONG TAIL- SOMETIMES 24 TO 26 FEET LONG.
CHICKS DO NOT RECOGNIZE WATER "INSTINCTIVELY." EVEN WHEN THIRSTY THEY MUST BE TAUGHT OR LEARN BY ACCIDENT OR BY WATCHING AND IMITATING OTHER CHICKS
?
"SHE CLAIMS IT'S NATURAL, BUT...."
★AN INDIANA HEN COMPLETELY CHANGED COLOR. BORN WHITE, SHE HAD A FEW BLACK FEATHERS HER SECOND YEAR, TURNED COMPLETELY BLACK HER THIRD!
★ AS A LOVE TOKEN CZECHOSLOVAKIAN GIRLS GIVE HAND PAINTED HARD BOILED EGGS TO THEIR BOY FRIENDS.
1931

WHEN DINING AT THE HOME OF RICH MEN, IN CERTAIN SECTIONS OF CHINA, IT IS PROPER TO TOSS ALL CHICKEN BONES OVER ONE'S SHOULDER. THIS IS A COMPLIMENT TO THE HOST, SHOWING THE GUEST REALIZES HE IS WEALTHY AND HAS ENOUGH SERVANTS TO TIDY UP AFTER THE MEAL
IN ANCIENT CHINA 5000 YEARS AGO, THE ROOSTER WAS CALLED KAI, THE BIRD WHICH TELLS TIME.
WHILE DRESSING A HEN SENT HER BY A NEPHEW FROM MEDFORD, MRS. G. HARRIS OF OAK PARK, WASHINGTON, FOUND IN THE GIZZARD THE DIAMOND SHE LOST SIX YEARS AGO WHILE VISITING HER NEPHEW AT MEDFORD
IN CHINA, ONE OF THE PENALTIES F SELLING A HOUSE WITH A LEAKY RO WAS BOMBARDMENT WITH EGGS BY TOWNSMEN.
COCK-ON-A-DRUM IS A CHINESE SYMBOL OF PEACE — THE DRUM, FOR MUSTERING TROOPS, WAS LONG UNUSED AND BECAME A COCK'S ROOST
PICKLED EGGS IS AN ANCIENT BUT INTERESTIN CHINESE DISH. COATED WITH CLAY AND RICE HUSKS, THE EGGS ARE PRESERVED FOR A YEAR OR MORE AND SERVED RAW WITH GINGER.
LIVE CHICKENS ARE TIED TO THE BEDS OF THE SICK IN CHINA. IT IS THE BELIEF THAT AT DEATH THE HUMAN SOUL WILL BE TRANSFERRED INTO THE CHICKEN'S BODY AND, THUS, BE KEPT IN CIRCULATION
THE IRISH FREE STATE PENNY (DESIGNED BY AN ENGLISHMAN) SHOWS A HEN AND CHICKS, THE EARLY CHRISTIAN SYMBOL OF GOD'S CARE FOR HIS CHILDREN
KIMONO JACKETS!
AN INDIANA RESIDENT HAS INVENTED AND MARKETED A KIMONA JACKET FOR POORLY FEATHERED CHICKS.
A CHINESE SUPERSTITION HAS IT THAT AN EGG FOUND ON MELTING SNOW IS LAID BY A PHOENIX WE DED TO A DRAGON. IF SUCH AN EGG IS HATCHED, DIRE TIMES WILL COME ABOUT

Cock of the walk. *(Terminology of cockfighters. Also means someone is conceited.)*
Cocky as a Bantam rooster.
Don't put all your eggs in one basket. *(Dates back to when eggs were carried in baskets.)*
Better an egg today, than a hen tomorrow.
He that buys eggs, buys many shells. *(Life is fragile, uncertain, good and nourishing, but perishable.)*
If a cock crows more than usual it's a sign a stranger is coming.
If a cock crows on going to bed, he's sure to rise with a watery head. *(Prophecy about wet weather.)*
A cock is worth most on his own dunghill. *(Latin proverb.)*
Fat hen lay no eggs. *(Old proverb.)*
One chick keeps a hen busy.
The hen that cackles is the one that has laid.
Flew the coop. *(Gone.)*
Hen scratching. *(Bad writing.)*
Good egg. *(Good person.)*
Bad egg. *(Bad person.)*
Got his hackles up. *(Angry.)*
Tough old rooster. *(Tough man.)*
His comb is beginning to turn red. *(A boy old enough to notice girls, as a cockerel begins to notice pullets when his comb begins to redden.)*
Something to crow about. *(Something to boast about.)*
Nothing to crow about. *(Nothing to boast about.)*
As hard to get along with as a sitting hen. *(A person with a bad personality, easily angered.)*
Chicken-hearted. *(This is meant to be soft-hearted or faint-hearted. However, chickens, especially game chickens, are brave, stout-hearted and persevering.)*
Hard- boiled. *(Tough.)*
Scratch for a living. *(Working hard, generally under harsh conditions, to make a meager living.)*
His feathers fell. *(A disappointment.)*
Feather his nest. *(To make his situation more comfortable at someone else's expense, generally money.)*

LIVE CHICKENS ARE SEALED UP IN SOME BUDDHIST IMAGES AT TIME OF CONSECRATION. THE LIFE OF THE CHICKEN IS THOUGHT TO PASS INTO THE IMAGE AND TRANSFORM IT INTO A LIVING BEING
GOSH, THAT'S A LOT OF EGGS TO EAT!
THE ANNUAL CONSUMPTION OF EGGS IN THE UNITED STATES IS 32 BILLION– AN AVERAGE OF ALMOST 270 PER PERSON
INTERNATIONAL SPIES HAVE MADE INGENIOUS USE OF HARD-BOILED EGGS. SECRET MESSAGES, WRITTEN ON THE SHELL WITH SUGAR AND ALUM LEAVE NO TRACE, BUT WHEN THE EGG IS BROKEN THE LETTERING DISTINCTLY SHOWS UP ON THE EGG WHITE IN BROWN COLOR
THE CHICKEN BREATHES WITH ITS BONES!
UNHATCHED DUCKLINGS SO-CALLED "BALUTS" ARE A PHILIPPINO DELICACY. THEY ARE EATEN ABOUT TEN DAYS BEFORE HATCHING TIME SOME HATCHERIES PRODUCE 10,000 BALUTS EVERY WEEK, THE EGGS ARE HATCHED IN SACKS IN THE SUN.
–BORNE
A BARRED PLYMOUTH ROCK HEN WAS FIRST BARRED; AFTER THE FIRST MOLT SHE TURNED WHITE, THEN BECAME BARRED AGAIN.
IN TURKEY THE TURKEY IS CALLED THE AMERICAN BIRD
I MUST HAVE BEEN DRINKING INK!
A HEN IN SOUTHWESTERN NORWAY LAYS EGGS WITH BLACK WHITES!
A THIMBLE WAS FOUND IN AN OTHERWISE PERFECT EGG!
12 EGGS A DAY KEPT DEATH AWAY 116 YEARS. MRS. DORA MEYERS, JERSEY CITY, N. J., WHO DIED AT THE AGE OF 116 YEARS, ATTRIBUTED HER LONG LIFE TO EATING A DOZEN EGGS EVERY DAY SINCE HER EARLY TEENS.
MUSCOVIES, CROSSED WITH OTHER DUCKS, PRODUCE ONLY STERILE OFFSPRING. THE FRENCH, WITH WHOM THIS CROSS IS POPULAR, CALL THEM FITTINGLY "CANARD MULET" MULE DUCK!
★A SOUTH CAROLINA HEN LAID AND HATCHED HER EGGS IN A CROW'S NEST THIRTY FEET ABOVE THE GROUND.

CHARLOTTE, EMPRESS
MEXICO, FOR FEAR
OF BEING POISONED,
ATE NOTHING BUT EGGS
LAID IN HER PRESENCE.
DIVORCE
A NEW YORK WOMAN DIVORCED HER HUSBAND BECAUSE HE WANTED SIX EGGS EVERY MORNING AND A KISS FOR EACH EGG
VAN WYCK OF ELIZABETH, N.J., FELT SO HENPECKED THAT HE MOVED INTO THE CHICKENCOOP AND LIVED THERE FOR MONTHS PRIOR TO HIS DIVORCE
GERMAN WOMEN BOOTLEG THEIR OWN EGGS INTO BEAUTY SHOPS.
BEING SERVED A TURKEY WITHOUT DRESSING, A WASHINGTON MAN CHOPPED UP THE TURKEY WITH A HATCHET AND CARVED HIMSELF A DIVORCE.
HONK
BETWEEN SUNSET AND SUNRISE, ANIMALS, INCLUDING POULTRY, MAY NOT GO ABOUT BEREA, OHIO, WITHOUT A TAIL LIGHT, ACCORDING TO A CITY ORDINANCE.
A DETROIT WOMAN DIVORCED HER HUSBAND BECAUSE HE KEPT HEN'S IN THE BEDROOM
MRS. V.G. BECK, OF OMAHA, NEB., IS ASKING FOR DIVORCE, BECAUSE HER HUSBAND BEAT HER UP WITH A SITTING HEN, THEN REVIVED HIS WIFE AND TOLD HER TO COOK THE CHICKEN WHICH HAD SUCCUMBED. SHE WAS BOUNCED AROUND SOME MORE WHEN SHE REFUSED. "WHO EVER HEARD OF EATING A SITTING HEN?" MRS. BECK ASKED THE JUDGE
A NEW YORK WOMAN DIVORCED HER HUSBAND: HE DEMANDED SIX EGGS EACH MORNING – AND A KISS FOR EACH EGG.
MRS. GENEVIEVE TITUS LOOKED WITH DISAPPROVAL UPON HER SISTER'S MARRIAGE. AFTER THE CEREMONY SHE THREW EGGS INSTEAD OF RICE AT THE YOUNG COUPLE!
H.C. POTTS, DENVER, COL., WAS SENTENCED TO DEATH FOR KILLING HIS WIFE BECAUSE SHE DID NOT COOK CHICKEN DINNERS OFTEN ENOUGH TO SUIT HIM.
BROWN EGGS WHICH TEND TO A ROUND SHAPE ARE DARKER IN SHELL COLOR
THIS EGG, THROWN FROM THE KITCHEN OF AN ARMY TRAIN, HIT A MINNESOTA MAN IN THE EYE, CAUSING PERMANENT INJURY. IT COST UNCLE SAM $4,339.00

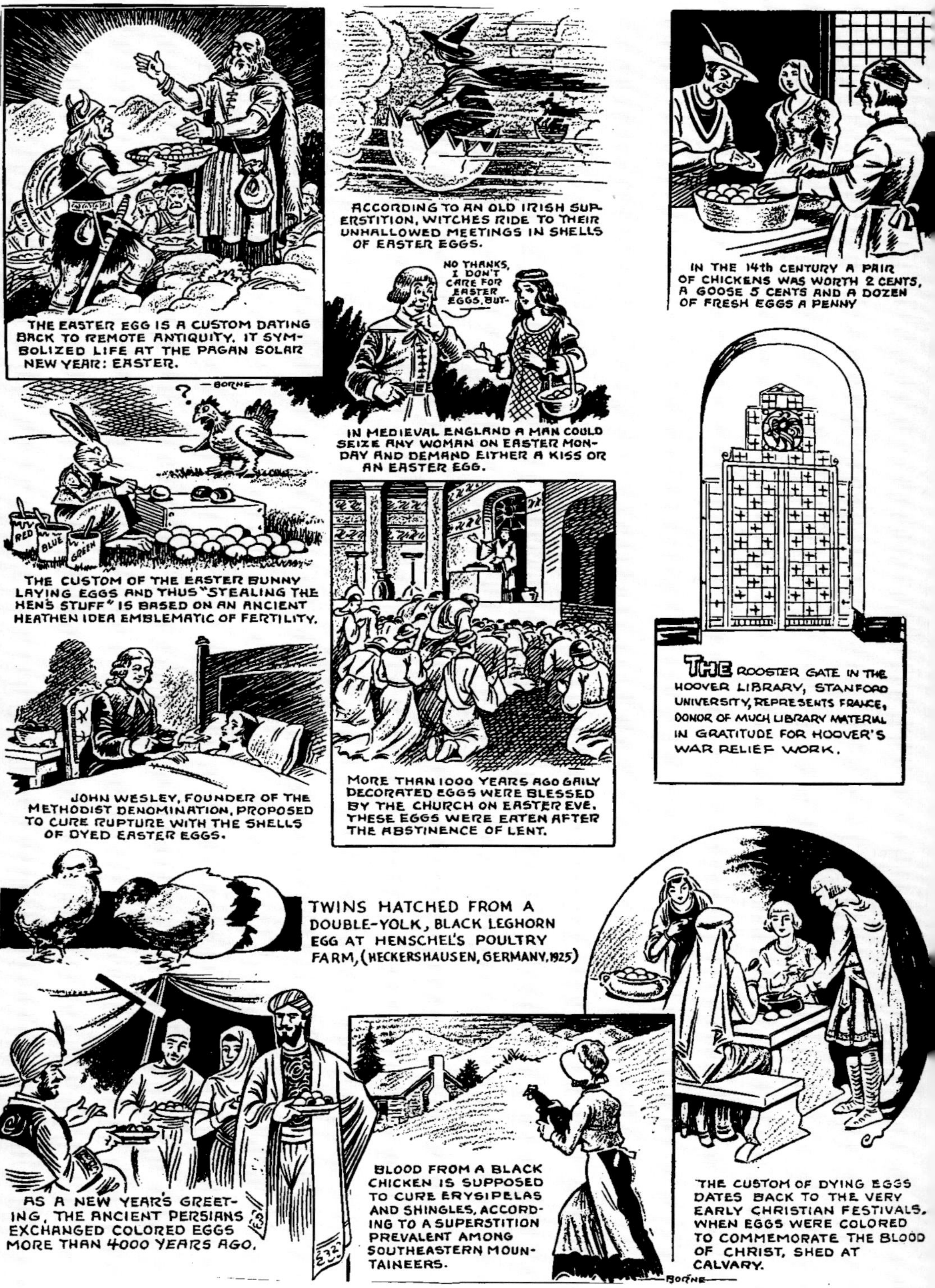

THE EASTER EGG IS A CUSTOM DATING BACK TO REMOTE ANTIQUITY. IT SYMBOLIZED LIFE AT THE PAGAN SOLAR NEW YEAR: EASTER.
—BORNE—
ACCORDING TO AN OLD IRISH SUPERSTITION, WITCHES RIDE TO THEIR UNHALLOWED MEETINGS IN SHELLS OF EASTER EGGS.
NO THANKS, I DON'T CARE FOR EASTER EGGS, BUT-
IN MEDIEVAL ENGLAND A MAN COULD SEIZE ANY WOMAN ON EASTER MONDAY AND DEMAND EITHER A KISS OR AN EASTER EGG.
IN THE 14th CENTURY A PAIR OF CHICKENS WAS WORTH 2 CENTS, A GOOSE 5 CENTS AND A DOZEN OF FRESH EGGS A PENNY
?
RED
BLUE
GREEN
THE CUSTOM OF THE EASTER BUNNY LAYING EGGS AND THUS "STEALING THE HEN'S STUFF" IS BASED ON AN ANCIENT HEATHEN IDEA EMBLEMATIC OF FERTILITY.
THE ROOSTER GATE IN THE HOOVER LIBRARY, STANFORD UNIVERSITY, REPRESENTS FRANCE, DONOR OF MUCH LIBRARY MATERIAL IN GRATITUDE FOR HOOVER'S WAR RELIEF WORK.
JOHN WESLEY, FOUNDER OF THE METHODIST DENOMINATION, PROPOSED TO CURE RUPTURE WITH THE SHELLS OF DYED EASTER EGGS.
MORE THAN 1000 YEARS AGO GAILY DECORATED EGGS WERE BLESSED BY THE CHURCH ON EASTER EVE. THESE EGGS WERE EATEN AFTER THE ABSTINENCE OF LENT.
TWINS HATCHED FROM A DOUBLE-YOLK, BLACK LEGHORN EGG AT HENSCHEL'S POULTRY FARM, (HECKERSHAUSEN, GERMANY, 1925)
AS A NEW YEAR'S GREETING, THE ANCIENT PERSIANS EXCHANGED COLORED EGGS MORE THAN 4000 YEARS AGO.
BLOOD FROM A BLACK CHICKEN IS SUPPOSED TO CURE ERYSIPELAS AND SHINGLES, ACCORDING TO A SUPERSTITION PREVALENT AMONG SOUTHEASTERN MOUNTAINEERS.
THE CUSTOM OF DYING EGGS DATES BACK TO THE VERY EARLY CHRISTIAN FESTIVALS, WHEN EGGS WERE COLORED TO COMMEMORATE THE BLOOD OF CHRIST, SHED AT CALVARY.
—BORNE—

Nest egg. *(A monetary saving.)*

To leave a nest egg. *(Leave something and not use all of one's money. When robbing a hidden nest, leave an egg in the nest so the hen will return and lay more. To play it safe, leave a nest egg.)*

Don't count your chickens before they hatch. *(This is the fable of a woman who began figuring how much money she would get for her eggs. With this money, she would buy things and continue to make a profit. She was still figuring her gains when she accidentally tipped over the basket of eggs and broke them all.)*

Feather in his cap. *(The American Indian who added a new feather to his headgear for every enemy slain, was the originator of this phrase.)*

Shakespeare wrote, *I will play the Swan and die in music. Mine honest friend, will you take eggs for money?* and *The cock, that is the trumpet to the morn, doth with his lofty and shrill sounding throat, aware the day.*

Chaucer wrote, *She hadde a cok, hight Chanticleer, in all the land of crowing nas his peer.*

1579 quotation: *I would not have him to count his chickens so soon before they are hatched.*

1580 quotation: *Children and chickens would ever be eating.*

1623 quotation: *The chicken is the country's, but the city eats it.*

1732 quotation: *It is no good a hen that cackles in your house, to lay in anothers.*

Irish quotation: *Proud of his plumage and his spurs, the feathered coxcomb struts, gallant and blithe, as an beardless cornet of dragoons.*

Colorful chicken trademark on California oranges

The following is being quoted with the permission of *Book Sales Company,* Secaucus, New Jersey.

Wild-goose Chase

How did a wild goose chase get that name?

A *wild-goose chase* was once a sort of game- a horse race in which the second and each succeeding horse had to follow the leader accurately and at a definite interval. Since the horses had to keep their positions like geese in flight, the chase was called a *wild-goose chase.* Since this was no race for no one could win - we adopted the phrase to describe a person following a course that leads to no goal.

Talking Turkey

Why do we say when a person gets down to the point that he's talking turkey?

It's an allusion to an anecdote. An Indian and a white man went hunting and agreed that they would share the game equally. They bagged three crows and two wild turkeys. The white man divided their bag and gave his hunting companion the first bird - a crow. Then he took a turkey for himself, gave the Indian a crow, took the other turkey and gave the Indian the last bird - a crow. When the Indian objected, the white man point out that the Indian had three birds to his two. To which the Indian replied, *We stop talk birds* we now *Talk Turkey.*

Turkey

How did our native American bird get the name turkey? (See Mexico)

Guinea hens and cocks were first imported into England from Africa by the way of Turkish dominions - and so were called *Turkey* cocks and hens. When the American bird was introduced it was confused with the African and given the same name. Then, when a distinction between the two birds was finally established and the names were differentiated, *turkey* was erroneously retained for the American bird instead of the African.

Lame Duck

Why is a member of Congress who has failed to be reelected called a lame duck?

Wild ducks in flight fly together - their head outstretched in front, their legs outstretched behind. A lame duck can't keep up with the flock. A member of the British Stock Exchange who couldn't meet his liabilities on settling day couldn't keep up with the flock and was struck off the list. From this we get our use of the term in reference to members of Congress who fail to be reelected and don't continue in office.

Chicken-Hearted

Why is a chicken considered symbolic of timidity in the phrase chicken-hearted?

The reference is to baby chicks. They are very timid and run to hide under their mother's wing on the slightest provocation.

Mother Carey's Chickens

Where did the birds known as Mother Carey's Chickens get that name?

The French speak of these birds *as the birds of our Mother St. Mary.* Another phrase applied to them is the Latin form for *dear mother* - mater cara. This sounds like *Mother Carey* and was so translated by British sailors.

White Feather

Why do we speak of a cowardly deed as showing the white feather?

When gamecocks are crossbred, a different colored *feather* shows up in their tails. This, more often than not, is *white.* It is generally believed that any outside strain will lead to cowardice in a gamecock - and so an act of cowardice is called *showing the white feather.*

China
The year of the rooster

Dutch Colloquialisms

By Hans Schippers

Such an egg, so the chicken. *(Little strokes fell great oaks.)*

Where do you find a Millers rooster who does not steal a little bit of grain. *(Opportunity makes the thief.)*

So, we do not part, said the cock to the worm and he ate her. *(Delay can be a danger in life.)*

A Brabant sheep, a Gueldre ox, a Flemish capon and a Frisian cow. *(The best of the best.)*

Everything with friendship!, said John and he took the eggs from his neighbors poultry nest. *(A hypocrite man or thief.)*

By the Comb and spurs we know the rooster. *(Clothes make the man.)*

What the rooster crows, chatter the hen. *(Women mostly talk like their husband.)*

If the rooster produces eggs, you will have them fresh every day. *(An impossible situation.)*

If the rooster crows on other hours than early in the morning, weather will change. *(When a man talks(crows) other than normal, something will change.)*

When the hen is angry she does not spare the rooster. *(Woman against man.)*

When he speaks, the rooster at the top of church tower starts to cry. *(Scoffing at a cocky speaker.)*

Do not believe it's day again, when you hear the crow of one cock. *(Do not be credulous.)*

It's not necessary to have two roosters in your backyard. *(Don't have two captains on a ship.)*

No rooster will crow to it. *(Nobody will notice it.)*

That's the talk of chickens. *(Nobody is listening to it.)*

A good rooster does not eat and drink without his hens. *(A good host pays attention to his guest.)*

A good rooster is not fat. *(A man who gives his wife what she needs is not thick.)*

It's for safety, said Dr. Lecoq and he took a large knife against a Bantam. *(Overdone measures.)*

IN THE SPREEWALD DISTRICT OF GERMANY- THE NORTHERN VENICE- CHICKENS ARE TAKEN TO AND FROM RANGE MORNINGS AND EVENINGS IN FLAT BOTTOMED BOATS.
WHEN A BABY IS NAMED IN BALI, THE HIGH PRIEST PUTS IT INTO A CHICKEN CRATE, SYMBOLIZING THE CHILD'S FIRST ACKNOWLEDGMENT OF HOME
EMPTY
A HEN WHICH LAYS EMPTY EGGS OF PERFECT SHAPE AND NORMAL SIZE IS OWNED BY MR. G. HUDDLESTON, KITTERY POINT, ME.
IN THE EARLY HISTORY OF POCKET WATCHES, THEY WERE CALLED "NUREMBERG EGGS", BECAUSE THEY WERE INVENTED IN NUREMBERG AND WERE EGG SHAPED,
IN PENRITH (ENGLAND) CHICKENS HAVE TO PASS OVER A POULTRY BRIDGE ABOVE A PUBLIC ROAD FROM THEIR HOUSES TO THEIR RANGE ON THE OPPOSITE SIDE OF THE STREET
ONE OF THE FEATS OF ANCIENT ROME CULINARY ART WAS TO SERVE FOWL ROASTED ON ONE SIDE AND BROILED ON THE OTHER-
PRIZE WINNER OF THE BURLINGTON LIARS CLUB: "MY WIFE IS SO LAZY SHE FEEDS THE CHICKENS POPCORN SO THE EGGS TURN THEMSELVES OVER IN THE FRYING-PAN."
HERE I GO AGAIN.
NEAR BLANCHESTER, OHIO, TWO TRUCKS CRASHED. ONE CARRIED EGGS, THE OTHER HAM.
HAM
EGGS
SEHEN SIE DAS ROOSTER?
A HUGE GOLD PLATED ROOSTER STANDS IN THE CITY HALL OF MUENSTER, GERMANY, MADE IN 1620 TO HONOR THE COCK THAT SAVED THE CITY. A BESIEGING ARMY GAVE UP AFTER 16 MONTHS WHEN THEY SAW A ROOSTER WALKING ON THE CITY WALLS, PROVING THAT THE CITIZENS WERE NOT STARVING!
AN OHIO HEN LAID AN EGG WITH A NICKEL IN IT. (A.P. DISPATCH TO DETROIT NEWS, 4/27-35
$2,500,000 FOR A DISH OF SCRAMBLED EGGS! A WAITRESS IN A LONDON CAFE SERVED T.J.THOMAS A PLATE OF SCRAMBLED EGGS WHICH HE ENJOYED SO MUCH THAT AT HIS DEATH THREE YEARS LATER, HE LEFT HER HIS FORTUNE.

He is walking like a rooster who stepped in the excrement's. *(He's walking like a fool.)*

I will make it short, said the fox and he was eating reluctantly an old rooster. *(The speaker was talking again about an old question.)*

Lean roosters crow the loudest. *(Stupid people cry the loudest.)*

Do not put eggs under the cocks, because they crack them. *(Do not entrust our business to somebody who is not familiar with it.)*

You cannot pluck good cocks. *(You cannot cheat a professional.)*

Divided cocks are for the fox. *(Discord breaks the power.)*

In early days the cock crows, said deaf John. Now they only gape. *(The man was more powerful in early days.)*

Go to bed with the chickens. *(Going very early to bed.)*

Before the hen picks the worm. *(Very quick.)*

She is like a boosted and spurred hen. *(She is like an Amazon, a robust girl.)*

As fresh as a hen. *(Fit and well.)*

So devoutly as a fox on his way to a hen-house. *(Roguish.)*

Being as silent as a hen with a cut throat. *(Being as silent as a grave.)*

Looking blue as a dying chicken. *(Looking really bad.)*

Falling as a fox in the middle of a hen-house. *(Coming unexpectedly, in a good or best situation.)*

No chicken to see. *(Nobody to see.)*

Seeing like a chicken. *(Bad eyesight.)*

Chickens crab backwards. *(A bad business without any profit.)*

A chicken-household. *(A mess.)*

Having a chicken heart. *(Being a coward.)*

Having a chicken life. *(A very comfortable life.)*

Being there as soon as the chickens. *(Being there as quick as possible.)*

Cackle like a chicken without a head. *(Incoherent talk.)*

VLADIMIR DUROV, FAMED RUSSIAN ANIMAL PSYCHOLOGIST, OPERATED A MINATURE TRAIN WITH A TURKEY AS CONDUCTOR AND A MONKEY AS ENGINEER AND FIREMAN.
THIEF CARTS OFF THE PROVERBIAL HOT STOVE WITH TWO CHICKENS INSIDE
THE CHICKEN DANCE IS A POPULAR ENTERTAINMENT AMONG THE NATIVES OF JAVA.
N.A. PAPAW, A RUSSIAN SCIENTIST, REMOVED THE LARGE LOBE OF THE BRAIN FROM 3-WEEKS OLD CHICKS AND RAISED THEM TO MATURITY BY ARTIFICIAL FEEDING.
TO PROVIDE MORE AND CHEAPER MEAT AND EGGS THAN PRODUCED BY POULTRY, RUSSIA IS PROMOTING EMU FARMS
WALLENSTEIN, DUKE OF FRIEDLAND-HE FAMED GERMAN GENERALISIMO OF THE THIRTY YEARS' WAR-WAS THE FIRST TO INTRODUCE COMMERCIAL POULTRY KEEPING AND EVELOP IT ON A COUNTRY-WIDE CALE. MOST OF THE PRINCIPLES ND FACTS CONTAINED IN HIS "RULES OF GOOD HUSBANDRY" PPLY TO THIS VERY DAY.
A PARROT OWNED BY MRS. A. E. WESTMEYER OF SAN ANTONIO, TEXAS, HATCHES AND MOTHERS CHICKS.
RUSSIA ERECTED A 6-FOOT STATUE OF A ROOSTER WHOSE FEMALE DESCENDANTS LAID MORE THAN 300,000 EGGS DURING THE PAST YEAR
IN SOME PARTS OF CHINA EACH PEASANT IS GIVEN A CHICK WHICH HE RAISES TO MARKET AGE AT HIS EXPENSE AND THEN TURNS IT IN AS TAX.
CLEMENT DOCKER
THE FAVORITE MEAL OF CLEMENT DOCKER .. 21 EGGS, 15 SLICES OF BACON, 2 LOAVES OF BREAD — ARE CARVED ON HIS TOMBSTONE AT BERKSWELL, ENG.
A HEN 28 YEARS OLD AND STILL AYING IS OWNED BY CHARLES W. GRAY

Farmer, watch your chickens. *(Be careful!)*

She cannot lay her eggs. *(She is not able to say what she wants to say.)*

Chicken manners. *(Caprices and whims.)*

Chicken, I have you. *(I've got it!)*

Chickens with feathers and all. *(Potatoes boiled in their skins.)*

He thinks he is a chicken but he's still in the egg. *(He thinks he reached the goal but he just started.)*

Looking like a just-born chicken. *(Looking very stupid.)*

A sleeping fox does not take a chicken. *(Do not fear the fools, let them sleep.)*

The hen who makes the most noise does not always lay the most eggs. *(A swaggerer.)*

A healthy sleeper goes to bed with the hens and wakes up when the dishes rattle. *(Loves to eat and sleep.)*

In the oats harvest the chickens are dear. *(As a hungry man eats, do not talk with him.)*

Watch the people who have no more fat on their forehead than a chicken. *(Do not do business with very poor people.)*

If the rich farmer eats chicken. he or the chicken is sick. *(Some farmers are so miserly that they only eat sick poultry.)*

The rooster crows the loudest in his own yard. *(At home he has much to say.)*

A lucky hatch, said John, twelve eggs and thirteen chicks. *(Very stupid bragging, impossible story.)*

From fried eggs never comes chickens. *(A very unusual story.)*

A little hen lays every day, an ostrich once a year. *(Do not discount the little ones.)*

As the egg breaks, it will stink. *(If a question is known by others it can cause problems.)*

N RURAL DISTRICTS IN FRANCE THE BRIDE
ILL STEPS ON AN EGG TO INSURE MARITAL
APPINESS.
ROM THE ENTRAILS OF THE CHICKEN, PRIESTS OF THE OLD ROMAN EMPIRE FORETOLD THE FUTURE. THEY WIELDED SUCH PUBLIC AND POLITICAL POWER THAT ONE CAN TRULY SAY:
"THE GUTS OF THE CHICKEN CONTROLLED THE DESTINIES OF THAT NATION"
© 1932
★IN CERTAIN PARTS OF HUNGARY, BRIDES DECORATE THEIR WEDDING CAKES BEAUTIFULLY, LEAVE THEM IN THE WINDOWS FOR A WHILE FOR PASSERSBY TO SEE, AND THEN FEED THEM TO THE CHICKENS.
A HEN ON AN OREGON FARM LAYS ONLY EGGS WITH DOUBLE YOLKS AND OF TREMENDOUS SIZE
THAT BIG ONE'S MINE
THIS ABBREVIATED COSTUME, MADE OF TURKEY FEATHERS, WAS WORN RECENTLY BY OREGON'S TURKEY QUEEN, MISS HELEN WETEL
-AND DO YOU KNOW THE HEN'S ADDRESS?
#2,500.00 FOR A DISH OF SCRAMBLED EGGS! A WAITRESS IN A LONDON CAFE SERVED T.J. THOMAS A PLATE OF SCRAMBLED EGGS WHICH HE ENJOYED SO MUCH THAT AT HIS DEATH 3 YEARS LATER HE LEFT HER HIS FORTUNE, MANSION AND AUTOMOBILES.
A SITTING HEN TURNS HER EGGS ON THE AVERAGE OF 120 TIMES A DAY!
A YOUNG MAN FROM NEW YORK TEACHES
★A MONASTARY IN GREECE HAS MAINTAINED A FLOCK OF ROOSTERS FOR ALMOST 1000 YEARS. NO HENS ARE ALLOWED

Chapter Ten

On the Lighter Side

Chicken Cartoons

The following cartoons are from George Burnham's book of 1855, *The History of Hen Fever.* They convey the exaggerated size of the relatively new breed called Cochins. Cochins zoomed into popularity both in England and the United States. When they were introduced into England in 1843 they were much larger than other chickens of that era. The Dorkings, Malay, Hamburgs and Polish were the dominate pure breeds. Large chickens such as Opringtons, had not yet been bred. In the United States, Plymouth Rocks, Rhode Island Reds, Wyandottes and Jersey Giants were not known in 1855.

An English Ambassador to China imported the first Cochins and presented them to Queen Victoria of England in December, 1843. This quickly made them known to the public and aroused an early interest in this massive new breed. The popularity of Cochins influenced the first poultry show in England in 1845 and the U.S. in 1849.

Guinness Book of Records

The following is quoted with permission of Guinness Books, Inc:

Chicken plucking: The fastest recorded time for plucking chickens was set in the 1975 contest at Masaryktown, Florida, on October 4, when a team of four women plucked 12 birds in 47.8 seconds. Leaving a single feather produces the cry *Fowl!*

The record for eating hard-boiled eggs, 14 of them in 58 seconds, went to Peter Dowdeswell in Corby, England, on February 18, 1977. The soft-boiled record, 32 in 130 seconds, was won by Douglas L. Burch, Korbets Restaurant, Mobile, Alabama, May 9, 1975. Eating eggs raw, 13 in 3.8 seconds, was the record for James Lindop, Manchester, England, on September 29, 1973.

The number of chickens eaten, 27 two pound pullets, was set by *Bozo* Miller, Trader Vic's, San Francisco, California, 1963.

A Rhode Island Red named *Penny* laid 20 eggs in seven days and seven in one day on September 11, 1971. It was owned by Treena White, Buckinghamshire, England.

The highest authenticated rate of egg-laying by a hen is 361 eggs in 364 days by a Black Orpington in an official test at Taranaki, New Zealand, in 1930.

The heaviest egg ever reported weighed 16 ounces and had a double yolk and double shell. It was laid by a White Leghorn, Vineland, New Jersey, February 25, 1956.

The largest egg recorded was one of almost 12 ounces and was a five-yolked egg. It was 12 and a quarter inches around the long axis and nine inches around the short axis. It was laid by a Black Minorca, Damsteads Farm, Lancashire, England, in 1896.

The highest claim for the number of yolks in a chicken egg is nine, reported by Diane Hainsworth, Hainsworth Poultry Farms, Mount Morris, New York, July, 1971.

Cartoons
Peter Gaymann
Fackelträger-Verlag
Hanover, Germany
1987

The highest weight claimed for a duck egg is 17 ounces and was laid by Cecilia, owned by Mrs. I. Grimsmo, Pine River, Minnesota, May, 1972.

The greatest live weight ever recorded for a turkey is 75 pounds. This was a holiday bird, reared by Signe Olsen, Salt Lake City, Utah, in December, 1973.

The record for a dressed bird is 71 pounds, 12 ounces and is for a stag reared in England, 1975.

The highest price paid for a turkey was $1,700 for 71 pound, 4 ounce, *Mr. Chukie* that was purchased by Leslie Grindle at the Waldorf Hotel, London, England, during an auction in December, 1976.

The heaviest chicken on record is *Weirdo,* a four year old White Sully weighing 22 pounds and reported in Calaveras County, California, in January 1973. *Weirdo* was so ferocious he crippled a dog and killed two cats.

The record flight by a chicken is 267 feet, eleven inches and was taken by *Mrs. Cluck* at a contest in Luckenbach, Texas, on June 8, 1975.

The longest feathers known are those of the cock birds of Japanese long-tailed fowls, or Onagdori which have been bred in southwestern Japan since the mid-17th century. In 1973, a tail covert measuring 34 feet, 9 and a half inches was reported by Masasha Kubota of Kochi, Shikoku.

Until the butcher will divorce us

Of Special Interest

I have a newspaper clipping about Eleanor Crzegorczyk, Bay County, Michigan, who lost her wedding rings after she peeled apples for a pie. Two weeks later, the rings were found in an egg her daughter used to make a cake. They had been eaten by a hen who enjoyed the apple peelings.

An egg nestled in the bosom of a lady for 21 days was hatched on a television program entitled, *Love of a Kind.* It was filmed in England in November, 1969. There was another incident in England where a similar scene was filmed by the British Broadcasting Company by the Earl of Snowdon, photographer-husband of Princess Margaret. This movie showed a middle-aged lady hatching an egg placed between her breasts. This was challenged by Sir Donald Gibson who claimed the temperature in her cleavage wasn't sufficient to hatch an egg. He felt extra clothing would be needed. However, in China, eggs have been hatched when held close to the stomach with layers of clothing.

Cartoons
Peter Gaymann
Fackelträger-Verlag
Hanover, Germany
1987

In 200 B.C. an Egyptian chemist discovered a unique method of cleaning pearls. He fed them to a chicken. The following morning, he searched through the droppings for the pearls. The claim is that this method restored the pearls to their beautiful luster. The original information pertaining to this is on rolls of papyrus and preserved in a Swedish museum.

In 1903, a patent was granted for chicken glasses. They were designed to protect the chickens eyes from being pecked. Today, commercially made specs for chickens are used to distract them and help correct the problem of cannibalism.

The *Chicken Dinner* candy bar was introduced in 1935 and cost only five cents. It did not prove to be popular even though the company launched an aggressive advertising campaign, which included a car designed in the shape of a chicken.

Researchers have found that hens converse with their eggs. A team of scientists at New York University determined communication between the hen and her eggs for 48 hours in advance of hatching. They identi-

fied 11 different auditory calls by which the hen and embryo communicated.

About 2300 years ago, Aristotle gave a good description of a chick's development and was the first to write about it. *After three days and three nights of incubation, the heart appears, connected with two vein ducts containing blood, and the heart begins beating. At ten days the embryo's head is larger than the rest of the body.*

Bantam is a province in West Java(Dutch East Indies), hence the name for our miniature chickens.

Egg yolk is one of the ancient binders used in tempera painting. Leonardo da Vinci's *The Last Supper* was painted with egg yolk paint.

In the United States *Blue Hen* was the nickname for Delaware. The Blue Hen of Delaware was one of many fighting chickens, not a domestic or exhibition breed.

There is a Persian saying that states four things may be learned from the cock; to fight, to get up early, to eat with your family, and to protect your spouse when she gets into trouble.

A National Geographic book, *Splendors of the Past*, states, *After the volcanic eruption that killed hundred in Pompeii, A.D. 79, an excavation revealed a bakery with 81 loaves of bread preserved. Also found were eggs, nuts and lentils. The bakery was totally sealed by the volcanic flow of lava.*

Misty Fagertrom, Oregon, had a Barred Rock hen that was chewed by a dog and nearly killed. It survived but with strange changes. The hen crowed and became very aggressive.

One hen was always glad to see her owner, but on Saturday, she would do a little *dance.* The basis of this unusual story is that a New England family always had beans and salt pork for supper on Saturday. The left-overs were given to the chickens. This hen knew a special treat was in the offing and would do her *dance* only on Saturdays, never at any other time.

In Greece, chickens were considered sacred and kept in temples. Early Greek coins had fighting cocks on them and cockfighting was popular.

Harry Kettle of Dallas, Texas, was the first to fly a chicken to a poultry show. This was in 1926 and the chicken was entered at the famous Madison Square Garden Show, New York.

In 1925, a poultry magazine exclusively for Buff breeds was published, called *The Golden Buffs*.

The World Poultry Congress of 1930 was held in the Crystal Palace of London and featured 20 acres under glass. It cost $8,000,000.00 and was a shopping mall with shops, libraries, restaurants, statues and relics of World War I.

Hans Schippers, Holland, wrote that only the rich in Europe had fine poultry. The poor stole from the rich, which is how the popularity of exhibition poultry spread in Europe. The stealing of exhibition poultry caught on in the US but it was not the poor stealing from the rich.

Cartoons
Peter Gaymann
Fackelträger-Verlag
Hanover, Germany
1987

In Pakistan, a leather technologist came up with the novel idea of utilizing chicken leg skin and processing it into leather. Because it is fairly strong, watch bands and ladies handbags were made of chicken leather. Though this was not done commercially, it is an interesting possibility.

Many people don't believe eggs will hatch if they are kept in a refrigerator. The fact is, they will, providing the temperature is not too severely cold and the eggs are turned regularly and not kept for more than two weeks.

In 1865, there was an English custom that said a gentleman upon entering English society would send a fat goose to the lady of the house.

Guineas start laying when Hickory trees bud.

"I'D SURE LIKE TO MEET THAT SLICK CHICK FROM POLAND!"

NO INCUBATORS IN ANCIENT DAY

"HE'S TRYING TO MAKE A GOOD IMPRESSION ON MISS AUSTRALIA!"

"I DON'T LIKE HIS 'GOD'S GIFT TO WOMEN' ATTITUDE EITHER BUT HE'S THE ONLY ROOSTER ON THE PLACE!"

Cartoons courtesy of Jim Zilverberg

The rooster is the national bird of France.

There is approximately one chicken for every person in the world.

In 1475, Peter Jesperz, Kennemerland, Holland, reported white poultry with black crests.

There is a small Alaskan village with a population of only 26, named Chicken. It is in this area of the state that the game bird, the Ptarmigan is prevalent. In view of this, the residents considered naming the town Ptarmigan (a member of the grouse family), but since none of them could spell it, let alone pronounce it, they called the town, Chicken.

Above the city of Juneau, Alaska, is an area called Chicken Ridge. The story goes that a chicken farmer had a difficult time and finally gave up in disgust. He decided to leave Juneau for good and released all his chickens to roam the area. Since then, the place has been known as Chicken Ridge.

The rooster weather vane from 1422 , was considered a Christian symbol of vigilance, Christ and the Devine Word. In 1484, Bishop Courtenay arranged for a copper cock weather vane to be placed on the spire of Exeter Cathedral. It was to remain there for 286 years. In 1752, it was taken down and erected 60 years later on the tower of St. Sidwells Church. It was taken down from there after 355 years and is probably in a museum.

'Tufted Duck'
by
Alfred Jacquemart
French
(1824-1896)
The Sladmore Gallery

Famous Quotes on Chickens

Jesus Christ:
How often would I have gathered thy children together, even as hen gathereth her chickens under her wings. Matthew 23:37

Winston Churchill:
When I warned that Britain would fight on alone whatever they (French) did, their generals said, In three weeks England will have her neck wrung like a chicken. Some chicken, some neck.

Uncle Remus:
Hongry rooster don't cackle w'en he fine a wum.

King Henry IV:
I want there to be no peasant in my realm so poor that he will not have a chicken in his pot every Sunday. France, 1553-1610

Don Marquis:
I have noticed that when chickens quit quarreling over their food they often find that there is enough for all of them. I wonder if it might be the same with the human race? (1878-1937)

George Elliott:
He was like a cock who thought the sun had risen to hear him crow.

Artemus Ward:
The sun has a right to set where it want to, and so, may I add, has a hen.

Anonymous:
The codfish lays ten thousand eggs, the homely hen lays one. The codfish never cackles to you when she's done. And so we scorn the codfish, while the humble hen we prize, which only goes to show that it pays to advertise.

Contests

Rooster Crowing Contest

The rooster crowing contest is perhaps the oldest such contest in the United States. It is sponsored by Rogue

River, Chamber of Commerce of Oregon and began in 1953. The idea was proposed by Shade Coombs, whom actually got the idea from Wales, England.

The Rogue River contest is part of a large celebration that includes a parade, Queen of the Rooster Crow, and other events. The contest is 30 minutes long, with a timing judge assigned to each rooster. Excellent prize money is offered and participants must send in their entry well in advance.

In 1958, a record 263 roosters were entered. The record crow was by a rooster named, *Beetle Baum* who vocalized with 112 cockle-doodle-do's in 30 minutes. No mechanical or electric device is permitted. Owners may give encouragement but they are not allowed to touch the bird or cage. You can imagine the fun the audience has watching the antics of the owner's trying to bring about the maximum of crows from the rooster. Imagine also, the frustration when a rooster will not crow at all.

The Goose with Shoes 'Andy' was hatched without feet. Gene Fleming of Hastings, Nebraska fitted Andy with childrens running shoes. With the help of these trendy running shoes, Andy can now walk, run, swim and stand still. Congratulations Andy and Gene!

Chicken Flying Contest

The International Chicken Flying Association, Ohio, is most unique. This yearly contest is sponsored by Bob Evan Restaurant, the Evan's farm near Rio Grande, Ohio. The first contest was held in 1971. Each entry has 30 seconds to fly from a ten foot high perch. Some chickens try, but do not fly at all, which is most embarrassing for the owner.

The greatest distance flown is a record 302 feet set in 1979 by Sherwood Costen, West Virginia. One thousand dollars is offered to break this record. The contest is held yearly on the third Saturday of May. A booklet of its history, rules, by-laws and the pecking order is worth sending for. Write to: International Chicken Flying Association, 3776 S. High Street, Box 7863, Station G, Columbus, Ohio, 43207.

Turkey Olympics

The Lake Waramaug Inn near New Preston, Connecticut sponsors the Turkey Olympics. It's a fun-filled day that includes weighing in, speed racing, high jump and the best dressed turkey. They even have a turkey parade. This event attracts spectators from around the area and is a popular social event.

Delmarva Chicken Festival

Delaware is a popular broiler producing state and holds the Delmarva Chicken Festival each year to promote their excellent broilers. The event is featured as the Olympics of the chicken world and includes an egg toss, egg spoon race, cackling hen contest, giant frying pan, chicken capers, rooster crowing contest and a *chicken scratch.* It is an exciting day that keeps Delmarva broilers in the consumers mind.

Duck Race

Deming, New Mexico is a small town that has gained national attention by sponsoring a duck race, which began in 1980. A whopping $7,500.00 in award money is offered. There are many events like a duck parade, the *best-dressed duck* contest, a *Duck Queen* is chosen and the *Great American Chili Cook-Off* is held. The Deming duck race is held in August and usually has over 300 entries with over ten thousand fans attending.

The winner of several races is a man with the name of Robert Duck, which is his real name. Mr. Duck is a jeweler from Albuquerque, New Mexico, and has been participating in the duck race since its inception. He has won a total of $30,000.00. Stromberg's has been supplying his ducks for several years.

Opposite
Chicken art from Siberia, made from end cuts of three kinds of wood, birch, pear and apple

To explain his exceptional wins, one must realize the diligent training sessions he has with his ducks and the care he takes in selecting the contestants, working with at least 100 ducks when he begins. The race distance is 16 feet long with a narrow two foot wide track, divided by a fence to control the ducks and minimize confusion. The timing is very precise, down to the split second and is observed by a qualified judge. The only monument in the US in honor of a duck, *Mary Quacker* is in Deming, New Mexico.

Chapter Eleven

Representation of Poultry by Country

Africa

Guineas

This wild fowl seems to have prevailed in most countries throughout the world and is referred to often in history. World Guinea authority, Hastings Belshaw, wrote that fossil remains of Guineas were found 2 1/2 million years ago in Czechoslovakia and dated back to the Pleistocene era.

Pliny, Varro and Columella were early writers who call the Guinea *Numidian.* Guineas were raised by the Roman farmers as early as the beginning of the Christian era. Aldrovandi, in his book of 1600, devoted five pages to Guineas, including a picture. In the 4th century a Guinea was featured in a mosaic which is in the National Museum in Rome. This fowl was also pictured in murals found in the Egyptian Pyramids.

Guineas were known to have existed in Mexico in 1524 and were later introduced into Spain. The English enjoyed Guineas after the Romans brought them to Britain. Colonists took the first birds to America. Worldwide distribution is evident because they were known to exist in New Zealand in 1864. In India, they were called Chinese Fowl.

While these birds have been domesticated for years, Guineas still retain their wild characteristic of being nervous and flighty. In the United States, they are found on farms and raised as a hobby. Guineas are alert and are able to detect intruders, which they do with their shrill cry.

Stromberg Hatchery advertised them in their catalog as the '*watchdog of the barnyard'.* They are recognized as a natural way of discouraging vermin like rats and snakes.

Guineas are not exhibited in poultry shows, with the possible exception of France. They are listed in the *French Poultry Standard,* but no other nation includes them. In France, Guineas are raised commercially in large numbers and merchandised extensively. I predict their popularity will increase in the U.S. because they are a delicacy.

Like pheasants, Guineas lay in late spring. Hatcheries which sell keets have a late season since they are not available until June. For more information on this fowl, I recommend the book, *Guinea Fowl of the World,* by R. H. Hastings, Belshaw, Ireland.

Ralph Winters, New Vienna, Iowa, is the largest breeder of different colors and anticipates offering them in Red and Chocolate. He now breeds Pearl Grey, White, Royal Purple, Coral Blue, Lavender, Porcelain, Buff and Slate.

Weights

Cock	4 pounds (1.80 kg)
Cockerel	3 1/2 pounds (1.60 kg)
Hen	3 1/2 pounds (1.60 kg)
Pullet	3 pounds (1.35 kg)

Above: Lavender
Middle: Pearl Grey
Below: Buff
Opposite: Coral Blue

Australia

Australian Pit Game Fowl

Australian Pit Game Fowl
Cock

I have no information to offer regarding the development of the Australian Pit Game Fowl except that they were originally developed from cockfighting stock. In Australia, however, they are not recognized as a fighter. They have specific breeding standards with characteristics described for exhibiting. The Australian Pit Games are shown frequently and are well established.

Weights

Cock	8 1/2 pounds (3.85 kg)
Cockerel	7 1/2 pounds (3.40 kg)
Hen	6 1/2 pounds (2.95 kg)
Pullet	5 1/2 pounds (2.50 kg)

Black Australorps

Black Australorps
Cock and hen

The *American Standard of Perfection* reports Black Australorps to be an egg production bred Black Orpingtons. This is an accurate conclusion. The *British Poultry Standard* refers to them as Australian Black Orpingtons. In Australia, the original Cook Black Orpingtons from England were crossed with Chinese Langshans. The *Standard* also mentions Black Cochins, Leghorns and Minorcas as the bloodlines introduced into their development. The long range goal was to produce a better layer.

Black Australorps were first exhibited in 1891 by Joseph Parington. In 1919, they were entered in egg laying contests where they performed very well. One of the best records was that of a hen that produced 338 eggs in one year. Breeders in the United States entered Black Australorps in egg laying contests and they made a good showing.

In Minnesota there was the Edmunds strain of Black Australorps which was the result of selective trapnest breeding. Many hatcheries used males to sire White Leghorn females to produce the popular Austra-White cross. With the development of commercial hybrid chickens, the Austra-Whites were phased out of the egg production scene.

In the U.S. Black Australorps are fairly popular for exhibiting and were accepted into the *American Standard of Perfection* in 1929.

Weights

Cock	8 1/2 pounds (3.85 kg)
Cockerel	7 1/2 pounds (3.40 kg)
Hen	6 1/2 pounds (2.95 kg)
Pullet	5 1/2 pounds (2.50 kg)

Elizabeth Duck

The Elizabeth Duck is relatively new in Australia. This breed was developed by Lance Ruting, New South Wales in 1972. They are a derivative of the Mallard or French Rouens. On the other side of the family there were only brown farmyard ducks. The Elizabeth duck was bred for quick development of meat. They are an attractive duck, and are bred and shown with specific characteristics described in the *Duck Standards.*

Trio of Elizabeth Ducks

Weights

Cock	7 pounds (3.20 kg)
Hen	5 pounds (2.25 kg)

Austria

Altsteirer

The Altsteirer was developed from ancient farm fowl north of Graz in the Stiermark region. They were mentioned in early poultry books and are known to have existed in the 14th century. In 1885, their breeding was systematically strengthened for egg laying and meat qualities. In 1894, breeding guide lines were established. The Altsteirer is recognized in the exhibiting Standards for Germany and Switzerland.

Weights
Cock 6.6 pounds (3.0 kg)
Hen 5.5 pounds (2.2 kg)

Above: Cock
Below: Hen

Opposite Above:
Antwerp Bearded Cock
Antwerp Bearded Hen

Below Left:
Antwerp Bearded Cock
Below Right:
Antwerp Bearded Hen

Belgium

Antwerp Bearded Bantams

In the United States, the Antwerp Bearded Bantams are referred to as the Antwerp Belgium Quail Bantam. The term '*quail*' denotes a color designation. In Belgium there are 20 colors of this fowl. Unlike many breeds, there is no larger size in this species.

The Antwerp Bearded Bantam is among the oldest known breeds and was first noted in the 17th century. Dutch artist, Albert Cuyp, produced a painting that places them between 1620 and 1691. The French painter, Rene Delin, painted the Cuckoo Belgium Antwerp Bearded Bantams. These birds are extremely popular. At a 1909 poultry show 6,890 Antwerp Bearded Bantams were exhibited by 961 breeders.

Weights

Cock	700 grams
Cockerel	600 grams
Hen	600 grams
Pullet	500 grams

Top
Brakel Cock

Middle
Brakel Hen

Bottom
Mechelner Hen

Brakel

The history of the Brakel has been lost. Sir Edward Brown wrote the Brakel was extensively bred in the neighborhood of Nederbraekel in Flanders, hence the name Brakel. Horst Schmidt states this breed was related to the Totleger and Mowe and came from old farm stock. M. Vander Snich suggested that some Belgian breeds along with others found in Germany, Italy, France and Spain, came to Belgium via Russia.

There are claims that Campines, known in the U.S. and England are one and the same chicken. There is no question that there is a relationship, but in Belgium, fanciers insist Campines are another breed, smaller than their Brakel.

Brakels were introduced into England in 1898. They are very good layers, most distinctive and attractive in their appearance.

Weights

Cock	6 pounds (2.70 kg)
Hen	5 pounds (2.25 kg)

Mechelner

There is nothing known of the origin of the Mechelner. Their name is derived from the town of Mechelen being bred in that vicinity. To increase their size Brahmas, Cochins and Belgian Game blood were used. The Mechelner was developed as a meat bird for markets in Brussels and Antwerp.

They were recognized in a breeding standard book in 1891 and by 1900 had become popular with a German specialty club which was established in 1905. The Mechelner has a quiet temperament and considered a fair layer of buff colored eggs. They are a popular exhibition bird and the French call a similar chicken, Malines.

Weights

Male	10 pounds (4.5 kg)
Female	8 pounds (4.5 kg)

Brazil

Muscovy Duck

White Muscovy Drake

The Muscoy Duck is one of the oldest known breeds. Originally from the wilds of South America, Brazil has been given credit as its homeland. This duck has been known by other names; Muscovite, Turkish, Peruvian, Guinea, Pato and by the French, Cairon and Barbary. They are still found in the wilds of Brazil, Columbia, Venezuela and Central America.

In the wild the Muscovy builds its nest in trees and they are excellent flyers. They are also known as the quackless duck. Though the Muscovy doesn't quack, it has an audible '*hiss, hiss*' and is considered a rugged bird.

The flavor of their dark meat tastes more like goose or wild duck. This duck does not cross with others. Their eggs require five weeks to hatch in comparison to four weeks for other breeds. Commercially, their eggs do not hatch as well in an incubator as other ducks. It's better to hatch them under a duck or hen. The Muscovy is bred in three colors, White, Blue and Dark. It was recognized in the *American Standard of Excellence* in 1874.

Pastel Muscovy Gander

Weights

Adult drake	12 pounds (5.45 kg)
Young drake	10 pounds (4.55 kg)
Adult duck	7 pounds (3.20 kg)
Young duck	6 pounds (2.70 kg)

Musician Fowl Cock

Musician Fowl

The Musician Fowl of Brazil, also known as Singers, is one of the six longest-crowing chickens in the world. Nothing specific is known regarding their origin. The color pattern is not uniform because priority is given to the quality and length of crow. Most roosters crow only for a couple of seconds, but the Musician Fowl can crow approximately eight to ten seconds. It is not a large chicken and weights are not available.

Bulgaria

Benkovski Geese

The Benkovski is the leading breed of geese in Bulgaria. They were created by crossing local geese with White Emden and Grey Toulouse in the Plovdiv district of Benkovsi, hence their name. They are considered an excellent meat goose, unique in the fact that they are able to develop large livers of 500 grams with forced feeding. This is a plus for the Benkovski because in Europe there is a demand by gourmet cooks for goose liver.

Benkovski Goose

Weights

Gander	22 pounds (10 kg)
Goose	16 1/2 pounds (7.50 kg)

Shumen Blacks

Shumen Blacks were created in the region of Shumen in northeast Bulgaria. There is no information on how they were developed. In addition to their black plumage, their beaks and legs are also black. The Shumen Blacks are considered fair layers of white eggs.

Shumen Blacks Pair

Weights

Cock	5 pounds (2.25 kg)
Hen	4 pounds (1.80 kg)

Stara Zagora Reds

The Stara Zagora Reds were developed by crossing Rhode Island Reds with the local Star Zagora. Their plumage is red and their beaks and legs are yellow. They are considered fairly good layers.

Weights

Cock	8 pounds (3.60 kg)
Hen	5 1/2 pounds (2.50 kg)

Stara Zagora Reds Pair

Canada

Canada Goose

Canada Goose

This handsome goose, commonly called the Canadian Honker, is a wild goose. Indians are credited with their domestication. Because they are wild, they fly, and most of them migrate south for the winter. Many breeders pinion them to prevent them from flying away.

Occasionally the migration pattern of Canada geese is broken as happened at the famous Mayo Clinic in Rochester, Minnesota. Feed was made readily available to the geese during all seasons, so they stayed where they were making the local lakes their permanent home. Perhaps many patients of the Mayo Clinic felt comfort in watching these magnificent geese. This breed was acknowledged in the *American Standard of Excellence* in 1874.

Weights

Adult gander	12 pounds (5.45 kg)
Young gander	10 pounds (4.55 kg)
Adult goose	10 pounds (4.55 kg)
Young goose	8 pounds (3.60 kg)

White Chantecler Cock

Chantecler

This is the only breed to be developed in Canada and was founded by Brother Wilfrid Chateaine, a monk in the Trapist Order, while he worked as the manager for the poultry breeding plant at the Quebec Agricultural Institute.

Partridge Chantecler
Cock

The first mating took place in 1908 and their name was taken from two French words meaning *'clear singing'* from the play, *Chanteler.* The Chantecler is a good layer of brown eggs and in the beginning became popular enough to warrant a club of their own. Through the years their popularity has dwindled.

The White Chantecler was recognized by the American Poultry Association in 1921. The Partridge Chantecler was developed in Edmonton, Alberta, Canada, and admitted into the Standard in 1935.

Weights	
Cock	8 1/2 pounds (3.85 kg)
Cockerel	7 1/2 pounds (3.40 kg)
Hen	6 1/2 pounds (2.95 kg)
Pullet	5 1/2 pounds (2.50 kg)

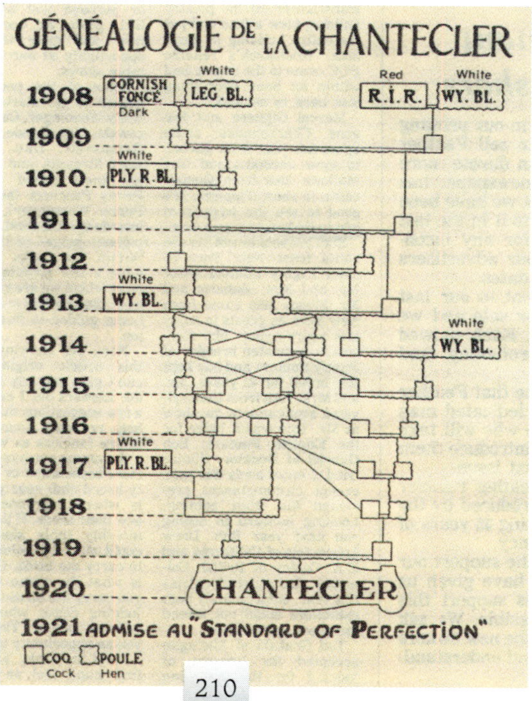

Chile

Araucanas

In the year 1560, missionaries made reference to Indians keeping chickens that laid colored eggs. Bonnington Moubray wrote of the chickens in South America. He stated the chickens laid colored eggs, their plumage was beautiful and they roosted in trees. Another reference from the year 1880, mentions the Araucan Indians having chickens that laid colored eggs. The Araucanas might well have been named because of their association with the Araucan Indians. Other researchers feel they were named after the Araucan forest of Chile. Studies have shown that chickens laying blue-green eggs were also known in Columbia, Ecuador, Brazil and Mexico.

Professor Salvador Castello, Spain, reported on the Araucanas at the first World Poultry Congress, September 1921, held at the Hague in Holland. He first observed them in 1914.

There were three types of Chilean fowl from which Araucanas evolved. Some had tufts, some were rumpless, some laid blue-green eggs, some laid brown eggs and some had a very strange way of crowing. In 1925, at the World Poultry Congress, Professor Castello gave another report on these unusual chickens. He

Araucana Lavender Bantam Hen

must have imported them because to the best of my knowledge there weren't any Araucanas in Spain at this time.

Dr. Bustos, of the Chilean army, mentioned that the native name for this unusual chicken was *'Quetros'.* He related the following story.

Araucana Cock

An Indian Chief Quaneano had numerous tail-less chickens. He preferred them because a fox could not catch them as easily as chickens with tails.

In further research I found the Araucanas were bred specifically to be tail-less, just for that reason.

Aldrovandi reports on a rumpless chicken, however, there is no statement that it laid blue-green eggs. There's a rumpless chicken that originated in Persia (Iran) so there could be some relationship.

The Secretary of the Araucanas Club in England wrote that there is a claim that tribes from Peking migrated to South America bringing the Araucanas. This might be a possibility since some chickens did originate in China. However, I know of no chicken in China with the genes to lay colored eggs.

Professor Uldoricia Prado, a naturalist, reported the blue egg characteristic did not appear in the chickens of the Araucan Indians until the 19th century. In Mexico, the Chachalaca known as the American Pheasant, was crossed with native hens to improve cockfighting stock. The resulting cross laid blue eggs.

There is another story that sailors returning on ships from South America saw chickens that laid blue-green eggs in Chile.

Dorian Roxburgh, British Araucanas Society, reported the first Araucanas came to Britain in the early 1920's via ships bringing fertilizer to Scotland.

In the United States, the first Araucanas were known to have been imported by Pratt Farms of Pennsylvania . In April of 1927, an issue of National Geographic mentioned the Araucanas and illustrated the breed with a fine painting. In September, 1948, National Geographic published an article entitled, *Easter Egg Chickens* and provided more information about the Araucanas.

Ward Bower imported his start of Araucanas in 1930 and it was from those and Pratt Farms stock that the present day Araucanas evolved.

Araucana Cock

The Stromberg Hatchery, Fort Dodge, Iowa , was the first large commercial hatchery to offer Araucanas chicks to the public. Many people wanted to exhibit this unusual chicken but the Araucanas weren't recognized by the American Poultry Association. They became even more controversial when breeders disagreed about the specification for showing. Adding to the problem was a story that Araucanas eggs had less cholesterol than other chicken eggs. Some stores sold these blue-green eggs at the fantastic price of $7.00 a dozen because of this falsehood. The food value of the Araucanas egg is the same as in all eggs.

It is true that Araucanas in the early days did not breed consistently true to color. My personal experience with them is interesting. Our breeder flock of over 200 birds looked like a bunch of mongrels with only a small percentage of rumpless chickens. Eventually, Araucanas breeders arrived at compromises to define color and split the group. Those to be shown with tails would be known as Americanas and the rumpless would be known as Araucanas. In England, they are divided into two classes, Araucanas and Rumpless Araucanas. All birds must have tufts.

Araucanas are also bred in Bantams. In the United States they may be shown in the following colors; Black, Black Red, Golden Duckwing, Silver and White. This breed was recognized by the American Poultry Association in 1976,

Weights

Cock	6 pounds (2.70 kg)
Cockerel	5 pounds (2.25 kg)
Hen	5 pounds (2.25 kg)
Pullet	4 pounds (1.80 kg)

Rumpless Araucana Pair by Mary Ellen Johnson

China

By Professor Xianpin Qiu

China is an ancient culture with a long history of poultry raising. When excavating in Wuan county, Hepei province along the Yellow River basin, grains and animal bones dating back 7,000 years were found. According to scientific identification by archaeologists, pieces of those bones belonged to chickens. Another finding was femoral bones in an ancient matriarchy tribe of 6,000 years ago, excavated in Banpocun, Xian Shanxi province. Although it has not been determined if the bones were from domestic or wild birds, it strongly supports existence of fowl in the Yellow River basin during the Stone Age.

In 1957, bones unearthed in dwelling relics in Shanmenxia of Henan province were identified to be chicken thigh and tibia bones, dated at 125 B.C. In 1954, in Jinshan county, Huepi province, a pottery chicken dated 2700-2650 B.C. was excavated. This evidence would indicate extensive raising of chickens at this time or chickens would not have been sculptured as a craft.

Evidence also exists about the raising of ducks and geese. Copper duck craft unearthed in an ancient grave in Pingquan county, Hepei province was dated around 2,000 years ago, B.C. In an ancient tomb of Yin-Dynasty (1711 to 1067 B.C.) ducks carved of jade were found. In the grave of a noble woman of 12th century cultural relics were unearthed. Among them was a carved jade goose. I conclude that duck and geese were being kept and were very popular during that period.

In 1898 to 1899, from the place of the Yin-Dynasty, Henan province, many tortoise and animal bones were excavated upon which divine oracles were inscribed. Writings show that chickens were quite popular about 3,500 years ago. It was told in lore that in the Yaoshun period (2500 B.C.) a minister named Beiyi was appointed to take charge of poultry raising in the country.

This building is named 'Fighting Duck Pool' within which a Chinese traditional drawing related to its name has been hung. The museum is located in Kun Shan country in the town of Jiao Zhi. There are several areas on the walls which are the only literal records to preserve the knowledge of this past sport. The restored Fighting Duck Pool existed in the Tang Dynasty. It is said that in the beginning of the West Han Dynasty (206 B.C. -23 A.D.) there was cockfighting, goose fighting as well as duck fighting. This was the entertainment for the nobility. The porches and railings made for a convenient spectator area. The long, ancient and historical Chinese River was the site of the duck fighting.

In the book of *Liezhi,* there was a story that Jishengzhi kept and trained fighting cocks for Emperor Zhou (823 to 782 B.C.). From 369 to 286 B.C., chickens specializing in fighting such as the Yanggou appeared and were described in the book, *Zhuangzhi.* The book, *Notes on Xijing* recorded that King Luguong (200 B.C.) was very interested in the game of cock, drake and gander fighting. Stories about feeding, breeding, management, classification of birds and geological distribution were colorful.

Statistical data taken from *Production Year Book of 1988* by FAO showed the total stock of poultry in China by the end of 1987 was 2,113 million head and included 1,796 million chickens, 316 million ducks and one million turkeys. This placed China as number one in the world. This is due not only to China's land, but also to the fact that this country owns colorful, native poultry breeds. They are seen in the countryside, city suburbs and plateaus as high as 3,200 feet above sea level.

Due to poultry raising, geological and ecological conditions in different areas of China, many native breeds have formed. In 1979 to 1982, a survey was conducted and each province compiled a book of poultry breeds that included all chickens, ducks and geese located in their province. Based on the survey, 52 of the prominent native breeds (including 27 chickens, 12 ducks, 13 geese , 9 developed breeds and 14 introduced strains) were compiled them into the book *Poultry Breeds in China.* I have selected 13 breeds from this book and one breed from *Fauna of Livestock and Poultry Breeds in Sichuan* to represent China.

African Geese

African Gander

For years these large geese have been known by the wrong name, which was called to my attention by Dr. Edmund Hoffman, Canada, who had been in China. He sent me a picture from China to verify what was their land of origin. The book by T.B. Miner, 1853, *The Domestic Poultry Book*, states that C.R. Belcher, Massachusetts, while reporting on the African, ties in the name Tchin Tchu, or China geese. They were also called Guinea Geese, Hong Kong, Poland, Chinese and Spanish Geese. They were eventually named the African because it was believed this name would have more appeal and the breed would sell better.

In 1852, African geese were imported into England and were recognized in the *British Poultry Standard.* African geese (Tchin Tchu and Tze Tau) were acknowledged in the *American Standard of Excellence* in 1874. White African geese were first developed in the U.S. by accident. From this start, Norman Kordosh and Dennis Hayzlett of Kansas perfected them over a ten year period.

Weights

Old gander	20 pounds (9.10 kg.)
Young gander	16 pounds (7.25 kg.)
Old goose	18 pounds (8.15 kg.)
Young goose	14 pounds (6.35 kg.)

Beijing Fatty

This breed originated in the suburbs of Beijing city because of great demand for chicken meat and eggs.

Economic quality: A dual purpose fowl with well developed egg and meat characteristics. Excellent as broilers with abundant meat flavor and rapid fattening.

Weights:

Cock	2,049 grams
Hen	1,730 grams

Shape and color: Comb, single, s-shape with unequal points, blade very small and thin. Crest, muff, beard and shank feathers seem to be a special characteristic of this breed. Small wattles. Profuse and soft-feathered. Body feather, brown showing brilliant shine. Tail feather, black in the cock. Tail, primaries and secondaries of hen, brown with irregular indication of black barring or feathering of half black and half brown that is separated by quill. Skin of all sections, beak and shank, yellow.

White Pekin Duck

Beijing Duck (Peking Duck)

As early as 1472, white lake ducks from Nanjing, Jiangsu province came to Beijing along the Beijing River. They took up residence in the suburbs of Beijing city and near by counties. A large number of ducks were needed to meet the need of consumers in the capital city. The Beijing was developed as a meat duck that took centuries of refining. These ducks were exported and were greatly welcomed because of their rapid growth rate, high meat qualities and high reproduction rate. The Beijing duck is well known for meat production.

Weights:

Day-old	63 to 65 grams
3 weeks	1,000 to 1,100 grams
7 weeks	2,700 to 2,800 grams

Shape and color: Color of beak, shank and toes, orange-red. Feather color, white.

Black Jiuyan

Black Jiuyan Cock

This chicken originated in Wanyuan county, Sichuan province, 3,000 feet above the sea level. This is a very cold mountainous area. Black Juiyuan chickens are known for laying blue shelled eggs. Analysis of egg qualities indicates higher breaking strength and thickness than other eggs from commercial breeds. Economic qualities: General purpose fowl.

Weights:

Cock	2,615 grams
Hen	1,762 grams

Shape and color: Comb, single or pea; Color of feather, black; Color of beak, shank, also black; Color of skin, black or white. In the white skin variety, color of comb, wattle and ear lobe, red. In the black skin variety, violet.

Chahua

The Chahua chicken is native to the Tai region of Xishuangbanla, Yunnan province. It is believed to have originated from Red Jungle Fowl.

Shape and color: Small in body size with close fitting feathers. It's a symmetrical, graceful, compact, active and fight-fancy chicken. Comb, single or pea, red in color. Beak, black or black with yellow. Color of skin, white or light yellow. Shank, black or black with yellow. Feather color of male: primaries, secondaries, main tails and sickles, black or red with black edge; red in other sections. Female, partridge brown in all sections except black in primaries, secondaries and tails.

Weights:

Male	1,190 grams
Female	1,000 grams

Chinese Game

Bird fighting had great popularity in China 2,700 years ago. After centuries of selection, a standard game chicken breed has come into being. The modern Chinese Game chicken shows erect carriage, relative length of shank, long and slim head and neck, long and trim body. According to their differences of geological distribution, the Chinese Game chicken can be classified into three types, large, medium and small.

Shape and color: Large type (Tulufan) was distributed in Tulufan, Shanshan and Tuoke counties of the Xinjiang region. Comb, pea; Color of comb, wattle and ear lobe, red. Color of beak, brown. Feather color, black, partridge or light brown. Color of skin, white. Shank, white or slate.

Weight:

Large	4,000 grams

The medium type (Chongyuan) was distributed in the extended areas of the Yellow River. Comb, pea; Wattle, small; Beak, brown, short, thick, and arched. Color of feather, black or red showing brightness. Shank, white.

Weight: Medium

Cock	3,870 grams
Hen	3,020 grams

The small type was developed in Xishuangbanla of Yunnan province. Comb, pea; Color of comb and ear lobe, red. Short, thick and arched beak. Color, yellowish-brown. Head, small. Feather color, black, white or brownish-red. Color of skin, white. Shank, yellow or slate.

Weight: Small

Cock	1,780 grams
Hen	1,120 grams

Chinese Geese

The greylag wild goose is recognized as the basis from which our geese evolved. There is a question if the Chinese geese were the first of the domesticated geese. The breed originated in the Lu-An Anhui province.

They were not widely accepted in Europe but were in the United States. In 1788, President George Washington imported the first geese into the U.S. Presumably, they were Chinese geese. This bird has also been called the Hong Kong, Asiatic, Swan, Guinea,

Left: Brown Chinese Gander

Right: White Chinese Gander

White African Pair

Spanish, Polish, Muscovy and Siberian geese. Chinese geese were recorded in northern India and Siberia according to Sir Edward Brown. White Chinese are the most popular although there are Brown Chinese geese. Chinese Geese were recognized in the *American Standard of Excellence* in 1874.

Weights

Gander10-12 pounds (4.55-5.45 kg.)
Goose 8-10 pounds (3.60-4.55 kg.)

The Cochin Story

Father Tachard *(or Sicard)* missionary and scientist, traveled throughout China in the early 1600's. He found large numbers of giant fowl, later to be known as Cochin chickens. Most of these were raised by members of the religious order Brahmo-Buddhists and followers of Khoung-foo-tsee and Leo-tsee, who worshipped ancestral spirits and believed in the immortality of all life forms.

Cochins were used primarily as food for the brotherhood and travelers. Surplus birds were sold to assist in the support of the monastery. None were used in religious ceremonies. Records from and old monastery retreat, Mi-aus, show the brotherhood raised this race of chicken for 1500 years. People sold or bartered the Cochins eggs and these large fowl were used for payment of taxes.

Reverend Jean Baptiste Prouix, a Belgium naturalist, journeyed to China in early the 1800's and was astonished by the size of these chickens. He observed the prevalence of the bird and noted that this chicken was '*very much a part of their religious tradition and superstition*'.

English navigator and explorer, Captain James Cook, (1728-1779) discovered the Hawaiian Islands. In 1778 he traveled to the island of Kauai where he found chickens. Unlike the English, apparently the natives did nothing to breed these birds with other fowl. In the middle 1800's native stocks of poultry were improved through the introduction of breeds from China, including Shanghaes, Cochin Chinas, Frizzles and Spanish.

White Cochin Cock

Because China and England were at war for many years, there was no communication between the two countries until the end of hostilities in 1843. Prior to that, the Chinese ports were closed to all foreigners. When the war ended, trade was re-established.

English ships returned with oriental goods, including tea and large chickens. Names given to the imported chickens reflected their place of origin. They were called China fowl, or specifically, Shanghae or Cochins, referring to the two areas of China where they came from. Shanghai is a port city of eastern China and Cochin is a region of South Vietnam on the south China

Buff Cochin Cock

Sea. English and American breeders favored the name Cochins.

The first Cochin Chinas arrived in England on December 23, 1843. Seven of these were given as a gift to Queen Victoria. She placed five females and two males in her deluxe poultry castle. Queen Victoria had a large collection of poultry and was the first to exhibit a pair of Cochins in April 1846, at the Dublin Cattle Show. Interest in the breed grew not only because of their beauty but because their docile nature made them easy to manage and confine.

Drawings by Harrison Weir depict Queen Victoria's first Cochin Chinas with clean shanks with some showing various degrees of feather density on the shanks. Artist renditions of poultry were the standard until the invention of photography in 1840.

The huge Cochins (12 to 13 pounds) were twice the size of most chickens in England. In addition to size, they were exceptional egg layers and not fussy eaters, all traits that made them desirable poultry to raise.

Feather coloration was not uniform, with shades of buff, reddish-buff, lemon and dark being common. English breeders wanted to establish uniform type and specific colors. Alfred Sturgeon was the first Cochins breeder of note in England. A ship returned from Shanghai, China, with eight very large, canary-colored chickens and in 1847, Sturgeon became the owner of these fowl.

It is believed three of them were accidentally killed as table birds and two died, leaving him with a trio of one cock and two females. The cock was a rich reddish-buff. One hen was reddish-cinnamon and the other a penciled brown. Nine chicks were hatched before the cock was killed by a fox.

Sturgeon's first Cochins were feather-shanked and Queen Victoria's were clean-shanked. It's not known to what extent the Queen's stock was used for breeding. Work done by Sturgeon credited him with being the father of *Cochin mania.*

Charles Punchard began his breeding activities in 1850 with partridge colored birds. His first exhibition was at the Great Exhibition Hall in London. Sturgeon's and Punchard's work was the beginning of the specialized breeding of Cochins. More breeders became interested in the species and Samuel Moody and Sam Not increased the breeding competition. Early English breeders instrumental in their development included Briggs, Bailey, Fortune, Peactor and Zurhorst. Their dedication contributed to the standardized breed type, color and fluffy characteristics.

Cochin popularity increased significantly. Restaurants capitalized on the interest by offering Shanghae dinners and soup. As prized as the Shanghae birds were with poultry fanciers, it's hard to believe they sold them for food. In all likelihood, many barnyard fowl were served as Shanghae specialties.

Barred (Cuckoo) Cochin

Cochins arrived in the United States in 1846. George P. Burnham imported two males and four females (clean-shanked) from Queen Victoria and paved the way for their popularity in the U.S. In December, 1852, Burnham sent a gift of seven pullet and two cockerel Cochins to Queen Victoria. In appreciation, she sent him a portrait of herself along with a written note of thanks, both a great honor to receive.

Burnham wrote two books, *The China Fowl - Shanghae, Cochin, and Brahma* written in 1877, and *The History of the Hen Fever*, which was the American counterpart to English *Cochin mania.* Many American breeders contributed to the perfection of the Cochins including Dr. Wright, Reverend Bumstead, Fassell and Cope.

Although most of the Cochins have feathered shanks,

Blue Cochin Hen

breeders in Japan have chosen to breed clean-shanked. They have been developed in a variety of colors, Buff, White, Partridge, Blue, Black, Cuckoo(Barred), Silver-Laced and Golden-Laced. One breeder in England is working on a Red Cochin and a Blue Breasted Partridge variety.

In modern China, there has been little change in the appearance of the Cochins since 1843. However, the fowl are referred to as Pu dong, Nine Catty Buffs and Beijing Oil chickens. They do not look like the Cochins known in other countries.

Cochins are distributed worldwide and interest in them is international in scope with breeders in the U.S., England, Germany, Holland, South Africa, Japan and Italy. For the hobbyist, there is an organization specializing in this species, called the Cochins International. For information contact Pat Lacey, W5055, Hwy 10, Neilsville, Wisconsin, 54456, USA.

White, Buff, Black and Partridge Cochins were accepted in the *American Standard of Excellence* in 1874. In 1965, Silver-Laced, Golden-Laced, Blue and Brown birds were recognized by the American Poultry Association in their *Standard.* In 1982, Barred (Cuckoo) was accepted in the *American Standard of Perfection.*

Weights

Cock	11 pounds (5.00 kg.)
Cockerel	9 pounds (3.20 kg.)
Hen	8 1/2 pounds (3.85 kg.)
Pullet	7 pounds (3.20 kg.)

Huiyuang Bearded

Originating in Huiyang, Gongdong province, this chicken is greatly welcomed in Hong Kong, Maco and other markets for its stout body shape, well developed breast muscle and early maturity.

Economic qualities: Small meat type.

Weight:

Cock	2,228 grams
Hen	1,601 grams

Huiyuang Bearded Cock

Shape and color: It has a striking appearance with yellow feather, beak and shanks and well developed beard, distinguishing it from all other breeds. It's called Three Yellow Bearded Chicken. Feather color of male; Sickle, golden yellow; Saddle and back, reddish yellow; Main tails, yellow with black at inside; Abdomen, buff. Feather color of female; yellow on all sections with a few black ones on primaries and secondaries.

Jingyuan

The Jingyuan developed in Jingling county, Gansu province and Guyuan county, Linxia Autonomous region, which is a plateau with average yearly temperatures of 6.7 to 8.3 C, 1,300 to 2,900 feet above sea level and relatively cold and dry. The Jinyuan chicken is a dual purpose fowl with well-developed egg and meat characteristics. The two varieties are named after their place of origin, Jingling and Guyuan.

Shape and color: Comb, rose; Color of comb and wattle, red. Feather color of male; primaries, secondaries and tail, black; Hackle, brownish-red; Breast and body sides; light brown; Abdomen, dark brown. Feather color of female: two varieties, buff and partridge brown in all sections except black in primaries, secondaries and tails.

Weight:

Cock (Jingling)	1,800 grams
Hen	1,630 grams
Cock (Guyuan)	2,250 grams
Hen	1,470 grams

Black Langshan Cock

Langshans

Black Langshans were imported into England by Major F.T. Croad in 1872, from the Langshan region of the Yangtze Kiang River. I found some Langshans were imported into California from China. I am sure most of the birds in the United States were from stock imported from England. There was a controversy as to whether the Langshans was a different breed or the same chicken as the Cochins. I have even found them referred to as Black Shanghai.

In Australia and Germany, Langshans are bred and exhibited

free of feathers on their shanks. Those with feathers are called Croad Langshans, though this reference doesn't apply in the U.S. In *Poultry Breeds of China*, three pair of black chickens are shown, all bare shanked. One pair is a Rose Comb , so the other two pair would be Langshans.

White Langshans came about as sports from the Blacks in 1885. Thomas F. McGrew. in his 1926, *The Poultry Book* reported on the Buff Langshans and Blue Langshans. However, the Buffs were not accepted because their appearance fell short. There are a few Blue Langshans in the U.S. but they are rarely exhibited.

For such a large fowl, the Langshans is a very prolific layer with a good rate of hatchability. Both the White and Black were accepted by the American Poultry Association in 1894.

Weights

Cock	9 pounds (4.10 kg.)
Cockerel	8 pounds (3.60 kg.)
Hen	7 pounds (3.20 kg.)
Pullet	6 pounds (2.70 kg.)

Above
White Langshan Hen

Below
Barred (Cuckoo) Langshan
German type,
no feathered shanks

Lindian

This chicken was named after its place of origin, Lindian county, Heilongjiang province. The average yearly temperature is 1.8 C, very cold and dry, and free of frost 125 to 130 days a year. The Lindian chicken is stout and capable of withstanding cold weather and coarse feeding.

Economic qualities: Dual purpose fowl with well-developed meat characteristics.

Shape and color: Comb, rose or single. Some chickens have crest, shank feathers, beard and muff. Color of skin, white; Beak, shank and toes, black. Feather color, two varieties, yellow and black. Feather of male, reddish-yellow with black tail.

Weight:

Cock	1,740 grams
Hen	1,270 grams

New Yangzhou

This breed originated from the crossing of New Hampshires with the native chicken flock in Yangzhou county, Jiangsu province. They were latter shipped to Zhejiang, Guangdong and Lianonin provinces.

Economic qualities: Bred for general purposes. Sexual maturity, 180 days. Average egg production within age of 500 days, was 197.6 eggs, average egg weight, 56 grams.

Shape and color: Comb, single, of median size. Head, moderate in shape. Beak, shank and feet, yellow. Feather color, dark or light red with some black feathers on primaries, secondaries and tails.

Weight:

Cock	3,290 grams
Hen	2,350 grams

Putong

The Putong or Nine Catty Buffs, the heritage of our modern Cochins. Note the light feathering on the shanks of this hen.

The east side of the Shanghai and Huangpu river is the place of origin of the Putong chickens. They feed on corn and beans as well as fish and sea food. Farmers prefer to keep chickens with large body size and lay large eggs. It was thought that the Cochins were developed from this chicken after it arrived in the United States. Because of its great body weight, yellow feather, shank and beak, it's also called Nine-jing Yellow.

Economic qualities: General purpose fowl with emphasis on production meat.

Shape and color: Comb, single and red, set firmly straight on head. Male has yellow chest with yellow back, red chest with red back, or black chest with red back. Tail color, black. Primaries and secondaries, partly black. Feather color of female, dark or light yellow on all sections with black edges around the webs. Hackle, primaries and secondaries, sometimes black. Has shank and toe feathers. Color of skin, shank and toes, yellow. Color of beak, yellow with brown front.

Weights:

Cock	3,550 grams
Hen	2,840 grams
Egg	57.9 grams

Putong Hen

Silkies

Though poultry books credit China as the source country of the Silkie, there is speculation that this information may not be accurate. A Chinese encyclopedia dated 1596, states that Silkies were also observed in Java, Sumatra, Ceylon, Ethiopia, Paraguay, Argentina, northern Asia and the Philippines.

Grey Silkie

Marco Polo first mentioned the Silkies in his travel logs and described them as black chickens that had *'fur like a cat'* and laid the best of eggs. Since Polo spent considerable time in China, the assumption has been made that he saw the Silkies there. However,, Polo did travel to other countries as well so there is some uncertainty as to whether or not he saw Silkies in China.

Hans J. Rignalda, Netherlands, is the author of a book on the Silkies *Zijdehoenders En Zijdehoenkrielen*, written in 1983.

He states, '*Buffon, 1635, in his book based on historia naturale wrote that he has seen Silkies on the Island of Madagascar and in Mozambique, Marsden, Sumatra and Indonesia. Inhabitants of Italy and Greece imported the Silkies from Ethiopia. The oldest description is from the Greek Aristotle';s of 380 B.C. He wrote about the Chinese Silkies. In Europe they were imported about A.D. 1600 - 1700. Gesner describes them in 1555 (white Ones). Ballas --- 1768 and 1773---Russia. For 4000 years the Mongols found Silkies in the Himalayas and took them north to China. They were common there between A.D. 1271 and 1295*'.

For years, especially in the United States and areas of Europe, the Silkies were referred to as Japanese Silkies, rather than Chinese Silkies. However, Noboru Suzuki of Japan reports that although the Japanese have made them a popular export chicken, the Silkies found in his country originally came from China via Chinese merchant ships. In Japan, the Silkies are called Uokkei and in China, Wugu-ji. Silkies were known in Asia and the interior of South America by the name Karaknath.

Silkies have been called Woolly Hens and in 1776, when they were exhibited in Brussels and Belgium, they were known as Rabbit Fowl. Due to the purplish-black mulberry color of their skin, face and wattles, the name Negro was also used. However, not every fowl referred to as Negro Fowl is a Silkie. The color of the bones of many of the Silkies led to another name, Black Boned chicken.

There are six sub-species of Silkies; Japanese, Siamese, African, Pink Skin, Buff and Tailless. Their appearance differs and the descriptions found are not precise.

Silkies have the following striking characteristics; Comb, strawberry, resembling the mulberry fruit, light violet before sexual maturity, red after sexual maturity. Crested head, large, well fitted on crown of head, composed of feather similar in texture to fine silk. Green ear lobe, bluish-green before sexual maturity, which disappears with time and becomes violet after sexual maturity. Beard and muff texture similar to crest. Wattle, small and similar to comb in color. Feather, fine, silky feathers except for primaries, secondaries and tail (uncompleted silky feathers). Five toes, some Silkies have a sixth toe. Shanks and fourth toe covered with feathers. Black pigment in bones, muscles, skin and internal organs.

Buff Silkie

Weights

Silkies (Jiangxi)

Cock	1,300 grams
Hen	970 grams

Silkies(Fujian)

Cock	1,810 grams
Hen	1,660 grams

Silkies may or may not have beards. Rignalda, of Holland, states bearded varieties are found in the United States, Australia and Holland. Color is a way of classifying the birds. The categories are, White, Black Partridge, Silver Partridge, Blue, Buff, Red, Lavender and Mottled. They were recognized in the *American Standard of Excellence* in 1874.

Weights

Cock	3 pounds (1.35 kg.)
Hen	2 pounds (.90 kg.)

Taihe Black Boned

Information on the health values of the Silkie Taihe chicken originated from an article in *The People Daily of China*, February 13, 1987, by Wei Rong. The following is a condensation of that article.

Black Boned chickens were raised in Taihe county, in the eastern Wushan Mountains in the Jiangxi province of China. A famous medicinal book written during the Ming dynasty entitled, *Compendium of Materia Medica* reports the Taihe Old Chicken is a tonic and nourishing food to treat women's diseases, to enrich the blood and build up health. It was especially honored by the Emperor Qian Long and Wushan Chicken. In 1911, the Taihe (Silkie) was exhibited at the International Fair of Panama and won a gold prize.

Research in China shows Taihe chickens contain certain hormones, blue pigment and amino acids which are required by the human body. These factors increase the blood cells and hemoglobin. Clinical experience has indicated that the Taihe chicken has effectiveness in treating women's sickness, sterility, menoxenia (abnormal menstruation) blood leuocorrhoes, metrorrhagia, tuberculosis, heart disease, neurasthenia and ostemomlacis in children.

Taihe eggs are an ideal nutritive, especially for old people and high blood pressure victims. Taihe chicken can be used as an ingredient for Chinese-made medicines like *Wuji Baifeng Wan* made famous by the Beijing Tongrentang Pharmacy using the Silkie chicken as a major source ingredient.

The Chinese Academy of Sciences showed that the medical and nutritional value of these chickens was due to natural spring water in the Wushan area where Wushan Farm was founded 30 years ago. They are raising over 100,000 Silkies a year to meet the demand. California raises a large number to meet Oriental consumption.

Taihe chicken is considered the treasure of the Taihe people. It makes exceptional tonic liquors like *Wuji Shenjijing*, made by refining the chicken, Ginseng and fruit of the Chinese Wolfberry. It takes up to ten months to make this liquor and it is a highly regarded gift.

In Asia, pills are made from the Silkie Bantam. The entire bird, including feathers, bones, blood and intestines are used in the manufacturing of these pills. They are considered effective for lung cancer, tuberculosis, menstrual symptoms and infertility. An ointment from liquefied feathers is used for external treatment. If a woman suffers from post-natal illness and exhaustion, she may get relief by eating the eggs of the Silkie. Another belief is that by consuming the eggs, arthritis of the legs may be controlled. Another by-product is used to treat children who wet their beds.

White Pekins

White Pekin

In China, White Pekins are known as the Beijing duck. They have been known since 1472 as the White Lake duck from Nanjing, Jiangso province. White Pekins had the Mallard as its heritage. Knowledge of their origin has been lost. We do know Pekins were imported to the United States in 1873 and to England in 1874.

The American duck meat industry strongly preferred White Pekins to their green ducks. White Pekins were acknowledged in the *American Standard of Excellence* in 1874

Weights

Drake	9 pounds (4.10 kg.)
Duck	8 pounds (3.60 kg.)

Wuding Hen

Wuding

Developed in Wuding and the nearby counties Yunnan province, the Wuding chicken is well known for large body size and excellent performance in meat production.

Economic qualities: Meat fowl for heavy meat production.

Shape and color: Comb, single, straight on the head. Color of comb, wattle and ear lobe, red. Beak and shank, black. Feather covered on shank and toes, profuse and soft feathering. Feather color of male, brownish red on all sections. Female color on primaries, secondaries and tail, black with crescent moon shape with white barring on body.

Xupu Geese

Originating in Xupu, Hunan province, this a typical Chinese White goose, belonging to a medium type.

Shape and color: Body, long, round, carriage very upright. Feather color, white. Color of skin, light yellow. Color of bill, knob, shank and toes, orange. Approximately 20 % of the geese in the original flock had a feathered crest on top of head.

Weights:

Cock	5,890 grams
Hen	5,330 grams

Yane Geese

Originating in Lu-an, Anjui province, Yane geese are a typical Chinese grey goose. It has compact feather covering and because of its large body size is considered a heavy breed.

Shape and color: Head, round shaped, with large erect rounded knob at front of skull, joining upper mandible. Leg and feet are short and black in color. Bill is medium length, stout at base. Color, black, some with throat wattle. Neck, long, slender, symmetrically arched. Body, round, plump, carriage upper right to show back angle of fifteen degrees. Feather color of gosling, dark green or brown. Color of bill, shank and toes, black. Feather color of matured geese, greyish-brown with a narrow greyish-yellow marking on the back side of the neck. Body feather, greyish-brown to greyish-white, changes from back to abdomen. Feather of wing, shoulder, legs, greyish-brown with white markings at edge. Knob and bill, black with half-rounded white feather marking at base. Toes, orange with black spots. Feet, black. Skin, greyish white.

Chinese Duck Fighting

The only physical evidence that the sport of duck fighting even existed was provided by Ching Chang Wang of Shanghai, China. He found pictures of a building called *Fighting Duck Pool* which was a museum area that recorded historical information. It's located in Kun Shan county in the town of Jiao Zhi. The town is split in two parts with bridges connecting them. In the scenic area that the restored Bao Sheng Temple is located the remains of the duck pool are evident. There are several areas on walls which are records preserving this past sport. I quote this condensation from the walls;

'Said to be the remains of the fighting duck pool which existed in the Tang dynasty. It is said that in the beginning of the West Han Dynasty (2-6 B.C. through A.D. 23) there was cockfighting, goose fighting as well as duck fighting. This was the entertainment for the nobility. The porches and railings make for a convenient spectator area'.

Wang wrote of a famous Chinese dictionary named, *Ci Hai*, (World of Ocean). It defines duck fighting as a game in ancient times to make ducks fight each other for the amusement of others. The historical Chinese River was the site of duck fighting and was alluded to in poetry.

Cuba

Cubalayas

The Cubalayas are the only breed to be developed in Cuba. The name is in honor of their land of origin. In the late 1800's, Domingo Montane had a breed called Malayos Pavos (Peacombed Malays) that was fighting stock from the Philippines. From Florida, Montane imported a Peacombed Malay pullet of a new color and the first breeding began. Later, imports arrived in Cuba from Europe and these were believed to be Spanish and Old English Games. These were crossed with the Malayos Pavos.

In 1939, the first Cubalayas were exhibited in the United States at the World Poultry Congress and International Baby Chick Association Convention held in Cleveland, Ohio. Cubalayas are an attractive and distinctive fowl. They are bred in White, Black and Black Breasted Red, with the Black Breasted Red being the most popular. This is a relatively rare breed with only a few being exhibited. All varieties of Cubalayas were accepted into the *American Standard of Perfection* in 1939.

Weights

Cock	6 pounds (2.70 kg.)
Cockerel	4 1/2 pounds (2.0 kg.)
Hen	4 pounds (1.80 kg.)
Pullet	3 pounds (1.35 kg.)

Pair of Cubalayas

Czechoslovakia

Bohemian Fowl

This native breed of Czechoslovakia was widely distributed in the middle of the 19th century. It was crossed with imported breeds. In 1913, a program was begun for their regeneration from the small population preserved in western Bohemia and Moravian highlands. In the development of the modern poultry industry, the numbers have declined. The Bohemian Fowl is a fair layer with a cream colored egg shell.

Weights

Cock	5 1/2 pounds (2.5 kg.)
Hen	5 pounds (2.2 kg.)

Bohemian Geese

Widely distributed throughout Bohemia, the Bohemian goose was not crossed with foreign geese until the 1870's. The larger foreign breeds gained in popularity so few of the true Bohemian geese remain. They are now bred only by the Bohemian Geese Club.

Weights

Gander	13 pounds (6 kg.)
Goose	11 pounds (5 kg.)

Top
Bohemian Cock

Middle
Bohemian Hen

Bottom
Bohemian Goose

Denmark

Danish Land Hen

The national breed of Denmark, the Danish Land Hen, is a very ancient breed believed to have originated during the Iron Age. Their long existence is proven by the half-burned hen bones which were found dating them at 2,000 years old. Their color reminds me of the Jungle Fowl of India. The Darwin Theory could establish them as one of our earliest chickens.

The Danish Luttehon, which has much shorter legs, is related to the Danish Land Hen. There may be a relationship with the German Creepers. This chicken is very active and considered a good layer.

Weights

Cock	5 pounds (2.25 kg.)
Hen	4 1/2 pounds (2.0 kg.)

Danish Land Hen Cock

Egypt

Egyptian Geese

Egyptian Goose

The Egyptian geese evolved from the wilds of Egypt and other nearby countries. It may have been the first domesticated fowl in advance of the first domesticated chicken. There is a painting from 2620 B.C., in the National Museum of Egypt, Cairo, that is known to be the first depiction of this fowl with the exception of cave paintings.

The Egyptians are the most colorful of all geese. They are very domineering geese compared to other varieties and are excellent flyers, which makes them interesting to keep with other poultry. Egyptian geese are exhibited only in the U.S. and were recognized in the *American Standard of Excellence* in 1874.

Weights

Adult gander	5 1/2 pounds (2.50 kg.)
Young gander	5 pounds (2.25 kg.)
Adult goose	4 1/2 pounds (2.00 kg.)
Young goose	4 pounds (1.80 kg.)

Fayoumi Cock

Fayoumi

The Fayoumi of Egypt has two other names, Begawi and Ramadi. Known to have existed in Egypt since 1880, they are named for the province of Fayoun where they are widespread. Though there is no knowledge of their origin, there is one theory that the Fayoumi is the progeny of the Brakel or Campines of Belgium and were brought to Egypt by

Napoleon. The second theory is that Fayoumi are the progeny of Asiatic chickens brought to Egypt by Senior Mohammed Ali Basha, President from the Anadol region of Turkey. He placed them in the Fayoum province. These chickens mixed with other native Egyptian fowl. I believe the second theory may be correct because of the presence of feathering on their shanks and toes which indicates a relationship to Asiatic genes.

Fayoumi Hen

One of the names, Begawi, is a town in the area of Begaa of Anadol, Turkey and the name Ramadi is associated with the region of El Ramad of the Foyoum province. The Fayoumi is not exhibited.

Weights

Cock	5 pounds (2.25 kg.)
Cockerel	4 pounds (1.80 kg.)
Hen	4 pounds (1.80 kg.)
Pullet	3 1/2 pounds (1.60 kg.)

Pharaoh Quail

Egypt made the quail popular all over the world. The Pharaoh quail is also known as the Coturnix in the United States and Japanese quail in Japan. It is possible they may also have come from nearby countries, but the Egyptians recorded them in history in various relief forms.

Pharaoh Quail

Popular for food, the Pharaoh quail is considered a special delicacy. The Chinese and Japanese have bred and improved them though they were known 900 years ago in China. The Pharaoh quail is known for its rapid rate of reproduction which enhances their popularity as a readily available food.

England

I was given inventory sheets from the Royal Archives of England of the Royal Aviaries, Frogmore, Windsor. The years covered are from 1896 to 1899 and 1910. They include an assortment of chickens, ducks, turkeys, guineas, pigeons, pheasants and even two ostrich. Surprisingly, there were no peacocks in the inventory nor were there Barred Rocks, Silver Laced Wyandottes, Golden Wyandottes or White Wyandottes. Many English Monarchs have been interested in poultry;

Queen Victoria	1840-1901
Queen Alexandria	1844-1925
George V	1865-1936
Edward VIII	1894-1972
George VI	1895-1952
Queen Elizabeth	1900-
Princess Anne	1950-

I think England is the only country with an Egg Exhibition Show. I visited the National Championship Show in 1989 and was surprised to see an excellent display of eggs. The *British Poultry Standard* lists exhibiting specifications for showing eggs and it is a competitive exhibition with awards.

Aylesbury

The Aylesbury duck is a very old breed. They were mentioned by Bonnington Moubray in his 1819 book *Practical Treatise on Domestic Poultry.* The Aylesbury is also referred to in an Encyclopedia of 1815. Sir Edward Brown states in his 1906, *Domestic Races of Poultry,'believed their start is from two white ducklings noticed in a group of wild ducks'.*

The White Pekins from China are known to have played a part in their

Aylesbury Drake

development. The Aylesbury originated in the Vale of Aylesbury, Buckinghamshire, thereby their name. They were bred for their meat qualities for the London markets. The Aylesbury duck was shown at the first English poultry show in 1848 and were recognized in the *American Standard of Excellence* in 1874.

Weights

Adult drake	9 pounds (4.10 kg.)
Young drake	8 pounds (3.60 kg.)
Adult duck	8 pounds (3.60 kg.)
Young duck	7 pounds (3.20 kg.)

Brecon Buff Geese

Brecon Geese

These beautiful geese were developed on the Breconshire Farms in Wales. I could not find any history on them. The Brecon Buffs are a relatively new breed and remind me of the American Buffs that were recognized in the *American Standard of Perfection* in 1947. The *British Poultry Standard* recognizes both the American Buffs and the Brecon Buffs.

Weights

Gander	19 pounds (8.60 kg.)
Goose	16 pounds (7.25 kg.)

Cornish

In a book by Edwin Mergargee published in 1926, there is reference to the fact that there were Cornish Games in Devonshire and Cornwall, England, as early as 1820. Fighting stock was bred by Sir Walter Raleigh Gilbert who crossed Red Aseels from India with English Black Breasted Games. The Cornish was composed of a Derby, Aseels and Black Sumatra cross.

Originally, the name Indian Game referred to Cornish. Sir Edward Brown presented a crossing that developed into what the United States called the Dark Cornish. Indian Games were first exhibited in 1870 in England. In the U.S., they were accepted as Dark Cornish in the *Standard of Excellence* in 1893.

Three varieties of Cornish were developed by American breeders. The White Cornish was accepted in the *American Standard* in 1898, the White Laced Red Cornish in 1909 and the Buff Cornish was recognized by the American Poultry Association in 1938. In the Standard the Cornish is referred to as Jubilee Indian Games.

Weights

Cock	9 pounds (4.10 kg.)
Cockerel	9 pounds (3.60 kg.)
Hen	8 pounds (3.60 kg.)
Pullet	6 1/2 pounds (2.95 kg.)

Top
Dark Cornish Hen

Bottom
Dark Cornish Cock

White Crested Duck

Crested Ducks

Paul Ives in his book, *Domestic Geese and Ducks* refers to Mr. Brent, who believes the Crested ducks are descendants of the Australian Tufted duck. He feels the rare Red Crested duck of New Zealand might be a possible heritage. It's obvious they are a very ancient breed. Jan Steen, an artist from Holland, shows two crested ducks in a painting done in 1660.

It is typical for some of the ducklings to hatch without crests. One breeder says they do not hatch as well as other ducks, due to their crests. They are an attractive and good meat duck. The Crested ducks were accepted in the *American Standard of Excellence* in 1874.

Weights

Adult drake	7 pounds (3.20 kg.)
Young drake	6 pounds (2.70 kg.)
Adult duck	6 pounds (2.70 kg.)
Young duck	5 pounds (2.25 kg.)

Dorkings

The Dorkings are definitely one of the oldest breeds. They are descended from fowl brought by the Romans during their domination of South Britain in 55 B.C. through A.D. 411. Columella, an early Roman writer of the first century, described a breed which was, in most respects, similar to the present day Dorkings. He referred to them as the purest breed that is five clawed. The fifth toe was mentioned by Pliny whose *History Naturalis* was published in A.D. 77. Pliny stated, '*Superiority of breed is indicated by an upright comb, sometimes double, black wings, ruddy visage and an odd number of toes*'.

Silver Grey Dorking Cock

Frenchman, De Reamure, also refers to a hen with five toes. Moubray , *Practical Treatise*, refers to them specifically as, *'Dorking fowl, so called from a town in Surrey, where they are to be found in great plenty and perfection, were probably first bred'.*

Aristotle mentioned the five toed breed and Aldrovandi has a picture in his book of a chicken with five toes. Reverend Edmund Dixon, in his 1853 book, *Ornamental and Domestic Fowls* mentions Dorkings being called Capon Fowl of England.

Silver Grey Dorking Hen

The Dorkings have changed much in type from the early days. In a drawing by Harrison Weir, the White Dorkings is more of a Leghorn type with a tuft of feathers in back of the comb. At one time there was a Rose Comb Silver Grey Dorkings. Their uniformity of type and color is a credit to English breeders.

There are seven color classifications of Dorkings, White, Speckled, Colored, Silver Grey, Black Breasted Red, Cuckoo and Red. England recognizes them in White, Silver Grey and Red. In the United States they are recognized in White, Colored and Silver Grey. Dorkings were accepted in the *American Standard of Excellence* in 1874.

Weights

Cock	9 pounds (4.10 kg.)
Cockerel	8 pounds (3.60 kg.)
Hen	7 pounds (3.20 kg.)
Pullet	6 pounds (2.70 kg.)

Pied English Call

English Calls

The British call these miniatures Decoy Ducks. In the United States they are known as English Calls. In the early days, they were referred to as the Dutch Dwarf Duck implying they came from Holland. The *British Yearbook of 1976* says the Chinese or Japanese were the earliest breeders. Their origin would be the Mallard and they were bred to supply sportsman with live decoys. They are bred in Grey, White, Brown, Pied, Blue Fawn, Silver, Smoky and Apricot. The White and Grey English Calls were recognized in *American Standard of Excellence* in 1874.

Weights

Adult drake	26 oz.
Young drake	22 oz.
Adult duck	20 oz.
Young duck	8 oz.

Khaki Campbell Pair

Khaki Campbell Ducks

The Khaki Campbell duck was introduced in 1901 by their originator, Adale Campbell of Uley, Gloucestershire. They were the result of crossing the Mallard with Indian Runners. The Khaki Campbell is known for their exceptional egg laying ability. A record egg production was established at 360 eggs in one year. They were recognized by the American Poultry Association in the *American Standard of Perfection* in 1941.

Weights

Adult drake	5 pounds (2.25 kg.)
Young drake	4 1/2 pounds (2.00 kg.)
Adult duck	4 1/2 pounds (2.00 kg.)
Young duck	4 pounds (1.80 kg.)

Modern Games

There is little information about the Modern Games. It is acknowledged that blood of the Malays played a part in their development. I think there is a relationship with Old English Games. In *Practical Treatise of Domestic Poultry*, Moubray mentions the Games as does Sir Edward Brown with Black Breasted Games being shown in 1851. Brown felt the pugnacity of these games indicated their relationship to fighting stock.

Bantam Modern Games were developed and became more popular than the larger species. They are bred in Black Breasted Red, Brown Red, Golden Duckwing, Silver Duckwing, Birchen, Red Pyle, White, Blue and Black. Modern Games were recognized in the *American Standard of Excellence* in 1874.

Silver Modern Game Bantam

Weights

Cock	6 pounds (2.70 kg.)
Cockerel	5 pounds (2.25 kg.)
Hen	4 1/2 pounds (2.00 kg.)
Pullet	4 pounds (1.80 kg.)

Old English Games

Though Old English Games are not considered a fighting chicken, their history is tied in with this sport. Julius Caesar wrote that the Britons kept fowls for pleasure and diversion, but not for table use, which would indicate cockfighting. Charles Darwin mentions their close resemblance to Jungle Fowl which could place them as one of the earliest breeds. Sir Edward Brown thought there was a possibility that a chicken associated with the Old English Game type may have prevailed in North and South America before the 15th century and that Spaniards may have brought them to Europe.

Red Pyle Modern Game Bantam

Bement, in his 1845 book, *American Poulterer's Companion*, uses an illustration very similar to Old English Games and refers to them as Irish Greys and Shawl-Necked. Harrison Weir has drawings of game

Golden Duckwing Modern Game Bantams

birds in his book that resemble Old English Games.

It is impossible to give an accurate history of their origin. Chickens of this type are all over the world, mostly as fighting stock. In the world of exhibition, the emphasis is on type and color. The *British Poultry Standard* recognizes over 30 colors.

In the United States the following colors were accepted in the *American Standard of Perfection* of 1938, Black Breasted Red, Brown Red, Golden Duckwing, Silver Duckwing, Red Pyle, White, Black, Spangled, Blue Breasted Red, Lemon Blue, Blue Golden Duckwing, Blue Silver Duckwing, and Blue Silver Duckwing. In 1965, the American Poultry Association recognized Blue Breasted Red, Lemon Blue, Blue Golden Duckwing and Self Blue.

Weights

Cock	6 pounds (2.70 kg.)
Cockerel	5 pounds (2.25 kg.)
Hen	5 pounds (2.25 kg.)
Pullet	4 pounds (1.80 kg.)

Black Orpington Cock

Orpingtons

Black Orpingtons were perfected by William Cook of Kent, England in 1880, who wanted to develop a new breed to merchandise. He wrote, *Poultry Keepers Account Book*, the most complete publication of its kind for that era and promoted them extensively. Developed originally as a utility fowl, Orpingtons quickly gained popularity for exhibiting and were shown in 1886 at the Crystal Dairy and Birmingham events. For a short period, there was a Rose Comb Black Orpingtons.

In 1887, the first Orpingtons Club was established. They were recognized in the *American Standard of Perfection* in 1905.

Buff

William Cook quickly developed and marketed Buff Orpingtons once the Blacks were well accepted. Breeding started in 1886 and the first appearance of Buff was in the 1894 Dairy Show. The Buffs became more popular than Blacks. The American Poultry Association recognized them three years before the Blacks, in 1905. A Buff Orpingtons Club was founded in 1898 and they were recognized in the *American Standard of Perfection* in 1902.

White

There is no history on the development of White Orpingtons. I have established that their origin had no relationship to other Orpingtons and that White Leghorns, White Cochins and White Dorkings were involved. It took several years of experimentation before the uniformity of White Orpingtons evolved. They were accepted in the *American Standard of Perfection* in 1905.

At the peak of the Orpingtons popularity other colors and varieties were developed, the Diamond Jubilee, Colombian, Cuckoo and Red. The latter did not survive the test of time. Blue Orpingtons were developed and accepted in the *American Standard of Perfection* in 1923.

Top to Bottom
Buff Orpington Cock
White Orpington Hen
Blue Orpington Hen

Weights

Cock	10 pounds (4.55 kg.)
Cockerel	8 1/2 pounds (3.85 kg.)
Hen	8 pounds (3.60 kg.)
Pullet	7 pounds (3.20 kg.)

Redcaps

In England, the Redcaps are closely associated with Derbyshire and sometimes known as Derbyshire Redcaps. They are also called Mooneys, Landshires, Yorkshires, Derbyshires, Rosetops and Yorkshire Ever Layers. Redcaps are closely related to Golden Spangled Hamburgs. In *Hick Poultry Encyclopedia*, it's mentioned, '*probably descended from the Hamburgs and English Game stock*'. They were recognized as a new breed in the *American Standard of Excellence* in 1888.

Weights

Cock	7 pounds (3.20 kg.)
Cockerel	6 pounds (2.70 kg.)
Hen	6 pounds (2.70 kg.)
Pullet	5 pounds (2.25 kg.)

Sebrights

Around 1810, Sir John Seabright was credited with the development of this Bantam breed. However, the creation transpired over several years. Sebright's accomplishment inspired an extraordinary

Above
English Redcap Hen

Right
English Redcap Cock

amount of interest in Bantams when the focus had been on larger fowl. In 1815 the first Sebrights Club was organized.

Sebrights are a special challenge to breeders because of their delicate lacing. They are bred primarily in Silver and Gold. Lemon Sebrights are being bred in Holland by Hans Schippers. This breed was accepted in the *American Standard of Excellence* in 1874.

Weights

Cock	26 oz.
Cockerel	22 oz.
Hen	22 oz.
Pullet	20 oz.

Silver Appleyard Ducks

Reginald Appleyard is credited with the development of this relatively new breed of duck. He started his work

Left
Golden Sebright Cock

Right
Gold Sebright Hen

Above
Crested Silver Appleyard

Below
Silver Appleyard

to perfect the breed in 1930 at the Priory Waterfowl Farm near Suffolk. His goal was to develop the ideal duck. His breeding remains a mystery and after Mr. Appleyard's death this breed required the support of the British Waterfowl Association to renew interest in them. Now, a miniature species has also been developed.

Weights

Drake	8-9 pounds (3.60-4.10 kg.)
Duck	7-8 pounds (3.20-3.60 kg.)

Sussex

The origin of the Sussex isn't known except that it is recognized as a very old breed and came from the areas of Kent and Sussex. They are closely related to the Dorkings. Sussex were very popular in the early days as a table fowl. They were also known as Surreys, Kents, Homestead Fowl and Sussex Barn Door Fowls. Early books such as Lewis Wright's of 1890, Tegetmeier's of 1867 and *The Poultry Book* by various authors in 1912, don't mention the Sussex though they discussed the early breeds.

Breeders became interested in establishing them after the first Poultry Show in England in 1853. The Sussex is included in the *British Poultry Standard* of 1865. In the *American Standard of Perfection*, the Speckled Sussex and the Red Sussex were recognized in 1914. The Light Sussex was recognized in 1929.

Weights

Cock	9 pounds (4.10 kg.)
Cockerel	7 1/2 pounds (3.40 kg.)
Hen	7 pounds (3.20 kg.)
Pullet	6 pounds (2.70 kg.)

Above
Sussex Hen

Below
Sussex Cock

Finland

Finnish Home Breed Chicken

The origin of the Finnish Home Breed chicken isn't known except that they date back to the 16th century. At one time there was an estimated 500,000, mostly in manors or large estates. From 1950 to 1960, they nearly became extinct with only 20 to 40 existing. Due to the dedicated effort of poultry fanciers, they now have numbers estimated at 1,000 to 2,000. Their preservation has been assured with official action safe-guarding their future.

The commercial strains of layers now dominates the poultry scene for egg consumption. The Finnish Home Breed chickens vary in color and are a tough bird that can endure the cold.

Weights

Cock	5 pounds (2.25 kg.)
Hen	4 pounds (1.80 kg.)

Finnish Home Breed Pair

France

Crevecoeurs

Crevecoeurs is one of the oldest French breeds. Precise origin is a mystery except that it evolved from the Polish. The distinctive two spike comb indicates other breeds were included in their original breeding. Sir Edward Brown suggests their area of origin was Normandy and Picardy, the new fowl being named after the village of Crevecoeur. They are an excellent dual purpose fowl that lays white eggs. The Crevecoeurs was recognized in the *American Standard of Excellence* in 1874.

Weights

Cock	8 pounds (3.60 kg.)
Cockerel	7 pounds (3.20 kg.)
Hen	7 pounds (3.20 kg.)
Pullet	6 pounds (2.70 kg.)

Avove
Crevecoeur Cock

Below
Crevecoeur Hen

Faverolles

The name of Faverolle is derived from the village of Faverolle, the area of their development. They were a multi-purpose breed with good table qualities and good brown egg production. They gained in popularity over the Houdans and were imported to England in 1886. Later, French Salmon and the rare White Faverolles were perfected and recognized in the *American Standard of Perfection* in 1914.

Weights

Cock	8 pounds (3.60 kg.)
Cockerel	7 pounds (3.20 kg.)
Hen	6 1/2 pounds (2.95 kg.)
Pullet	5 1/2 pounds (2.50 kg.)

Houdans

French Houdans is one of the older French breeds. They were developed from crossing Black Polish, Crevecoeurs and a common five toed fowl, likely Dorkings. They originated in the area of Houdan, France are bred for table qualities and are fairly good layers of white eggs.

At one time, the Houdans was France's leading market chicken. Introduced into England in 1850, they were first called the *Normandy Fowl.* Like most breeds, they have changed in appearance. An early painting presents them with white dominating their plumage.

The Houdans are exhibited at larger poultry shows. They were recognized in the

Top
Salmon Faverolle Cock

Middle
Salmon Faverolle Hen

Bottom
Houdan Cock

American Standard of Excellence in 1874. White Houdans were developed in the United States involving a cross with White Polish. They were admitted to the *American Standard of Perfection* in 1914.

Weights

Cock	8 pounds (3.60 kg.)
Cockerel	7 pounds (3.20 kg.)
Hen	6 1/2 pounds (2.95 kg.)
Pullet	5 1/2 pounds (2.50 kg.)

LaFleche

The LaFleche is named after the city of LaFleche in the valley of La Sarthe, north of Loire River where they have been bred for a long time. LaFleche, along with Crevecoeurs, have a distinctive two spike comb. They are exhibited in the United States, England, Germany and France. The LaFleche was recognized in the *American Standard of Excellence* in 1874.

Weights

Cock	8 pounds (3.60 kg.)
Cockerel	7 pounds (3.20 kg.)
Hen	6 1/2 pounds (2.95 kg.)
Pullet	5 1/2 pounds (2.50 kg.)

Top
Month old Houdan chicks

Middle
La Fleche Cock

Bottom
La Fleche Pair

Marans

The Marans are a popular French breed developed for their extremely dark brown eggs which sell for premium prices. There is a theory that the Marans were made up of Coucou de Malines, Coucou de Rennes, Croad Langshans, Faverolles, Barred Rocks and Gatinaises. Another theory is that they descended from small black fowl popular during the English occupation from the 17th century. There is even a third theory that says the Marans are an improved variety of the Coucou de Rennes. Cuckoo Marans are recognized in the *British Poultry Standard*. Other colors are being bred in France.

Weights

Cock	8 pounds (3.60 kg.)
Cockerel	7 pounds (3.20 kg.)
Hen	7 pounds (3.20 kg.)
Pullet	6 pounds (2.70 kg.)

Top
Maran Cock

Middle
Maran Hen

Rouen Ducks

This popular duck evolved from the Mallard. The Normandy area of France is considered the area of origin. How the name was derived is unknown. One version of the name is Rohan after a Cardinal. Another, Rhone, designating a province in France. To increase the size of the Mallard by three times took diligent and selective breeding of the Rouen duck. For many years this popular bird was sold to select Paris markets. Many were exported to England where they attained a high level of perfection.

Rouen Ducks
Pair

Weights

Adult drake	9 pounds (4.10 kg.)
Young drake	8 pounds (3.60 kg.)
Adult duck	8 pounds (3.60 kg.)
Young duck	7 pounds (3.20 kg.)

Toulouse Geese

The well known Toulouse Geese were developed from common grey farm geese in the area of Toulouse in Southern France. It is well known that the Wild Grelag Goose is the ancestor of all our geese. In France, they are known as Gers Goose and L'Ore de Gers Goose, Gers is a province of Toulouse. It is also known as the Mediterrean Goose and the Pyrennean Goose. The most recognized name is the Toulouse Goose. In England is was known as early as 1810. The first breeding standard was established by the London Poultry Club.

Toulouse Grey Pair

In the United States, the name which applies to the show type Toulouse Goose is the Giant Dewlaps. However, in exibiting the name Toulouse and Giant Dewlaps are synonomous. The Giant Dewlaps can attain weights of up to 40 lbs. A number of breeders have found it advantages to use artificial insemination with these geese because of their large size and clumsy nature. In the commercial sphere, the Grey Toulouse are not as large, are more productive, laying more eggs with a greater hatch rate.

The Grey Toulouse was recognized in the First Standard of Excellence in 1874. The Buff Toulouse was recognized in the American Standard of Excellence in 1977.

Weights

Gander	28-30 pounds (12.70 - 13.60 kg.)
Goose	20-22 pounds (9.10 - 10.00 kg.)

Augsburger Hen

Augsburger

The name Augsburger means the *fowl of Augsburg.* They were developed in 1870 by J. Meyer, Haunstetten, who produced a black type chicken with double comb, from a mixture of Lamott fowl and French LcFleche. Difficulty in purifying the stock caused low popularity until after 1945. H. Suttner rescued the breed by establishing their uniformity.

Weights

Cock	6.6 pounds (3.00 kg.)
Hen	5.5 pounds (2.50 kg.)

Bergische Kraher

The Bergische Kraher is considered the old national breed of Germany. They are also known as *Mountain Long Crowers.* Their history is not precise but it is believed that Spanish refugee monks brought them into the village of Wuppertal in the Bergishe land of Germany. Later, they crossed with local breeds known as Straw Hens. Another version of their origin indicates they resulted as sports from an egg producing breed called the Schlotterkamm. Their long-crowing ability is famous, crowing four to five times as long as an

Above
Bergische Kraher Cock

Right
Beilefelder Kennhuhner Cock

ordinary rooster. The Bergische Kraher was shown in 1853 and recognized in the *German Standard* in 1885. The first Specialty club was formed in 1916.

Weights

Cock	7.7 pounds (3.5 kg.)
Hen	5.5 pounds (2.5 kg.)

Beilefelder Kennhuhner Hen

Bielefelder Kennhuhner

This variety was developed from crossings in Germany that started in 1970. Auto sexing resulted in a dual purpose breed, producing large brown eggs. They were recognized as an established breed in 1980.

Weights

Cock	8.8 pounds (4.00 kg.)
Hen	7.1 pounds (3.24 kg.)

Dresdener

Dresdener Cock

The original breeder of the Dresdener fowl was A. Zumpe, whose aim was to develop a high vitality egg producing chicken with a Rose Comb, thus reducing the difficulty with frozen combs in cold temperatures. Zumpe started with Wyandottes, Rhode Island Reds and New Hampshires. By 1953, they were fairly uniform and could be shown.

The Dresdener was first called New Dresdens due to the rebuilding of their city. The breed was recognized in 1957 as a new breed. The Dresdener is considered a good layer and bred in White, Black, Rusty Jungle Fowl and Gold Brown.

Weights

Cock	6.6 pounds (3.0 kg.)
Hen	5.0 pounds (2.25 kg.)

Krupers

The name Creepers applies to the German name Krupers. This ancient type was introduced into Germany from Scotland in 1871. It is thought that perhaps the Krupers may have come from Denmark and France. They are a distinctive breed with very low carriage. They are bred in Black, White, Partridge and Cuckoo.

Weights

Cock	5.5 pounds (2.25 kg.)
Hen	4.4 pounds (2.00 kg.)

Krupers Pair
by
Mary Ellen Johnson

Niederrheiner

Niederrheiner Pair

J. Jobs imported North Holland Blues from Dutch farm stock in 1939. The stock was derived from Hachein Cuckoo, Barred Orpingtons, Barred Plymouth Rocks and Games and was called Blue Barred.
F. Regenstein assisted in perfecting this dual purpose stock.

In 1943, auto sexing was recognized, with the males being Blue Barred and the females Colored. In the 1960's the Blues and Birchens were acknowledged as a breed. They were especially noted for rapid growth and good egg laying ability.

Weights

Cock	8.8 pounds (4.0 kg.)
Hen	6.6 pounds (3.0 kg.)

Ostriesische Mowen

Ostriesische Mowen is from a Dutch and East Friesian old farm stock related to the Totlers and Brakels, and sometimes called the Campines.
There is nothing known about their development but they are good layers of white eggs, good foragers and tolerant of weather. The Ostriesische Mowen is bred in Silver and Gold.

Silver Ostriesische Mowen Hen

Weights

Cock	6.6 pounds (3.00 kg.)
Hen	5.5 pounds (2.50 kg.)

Pomeranian Gander

Pomeranian Geese

These attractive geese have been known as Saddlebacks and Mottled. There is no information about their crossing but they carry White Emden and Grey Toulouse bloodlines. The Pomeranian was developed as a commercially and is known for their excellent economic quality. They became very popular for exhibiting.

The *German Poultry Standard* does not differentiate, but in the *American Standard* there are Grey Saddleback and Buff Saddleback Pomeranians. In the *English Standard* they are referred to as the Grey Back. This variety was recognized in the *American Poultry Standard* in 1977.

Weights

Adult gander	17 pounds (7.7 kg.)
Young gander	15 pounds (6.80 kg.)
Adult goose	15 pounds (6.80 kg.)
Young goose	13 pounds (5.90 kg.)

Rheinlander Cock

Rheinlander

The Rheinlander was developed from local farm fowls. In the 1800's, Lemans and the Bergische Kraher were crossed. This bird is a dualpurpose breed and lays white eggs.

Weights

Cock	6.6 pounds (3.00 kg.)
Hen	5.5 pounds (2.50 kg.)

Sulmater

Mating of Sulmaters

The Sulmater originated from ordinary farm fowl in the area of Steiermark of Austria. From 1865 to 1875 German breeders crossed them with Cochins, Houdans and Dorkings to improve their size and produced the Steier, a meat fowl. In 1900, they were named the Sulmater.

The Sulmater was very popular, but after World War 1 they fell into decline and became very rare. This breed is a good meat production fowl and layer.

Weights

Cocks	8.8 pounds (4.00 kg.)
Hen	7.7 pounds (3.50 kg.)

Thuringer Barthuhner

Thuringer Barthuhners Pair

The name Thuringer Barthuhner means *The Bearded Fowls of Thuringia*. Bearded chickens were known by the middle of the 18th century in the region of Ruhla City, Thuringia. By 1880, there was a dozen varieties. The name Thuringian Chubby Cheeks and Otterheads of West Thuringia also applies to them. The original ancestry was probably Paduan fowl.

The Thuringer Barthuhner is a good forager and layer of white eggs. They are bred in Black, White, Blue, Yellow, Cuckoo, Gold, Silver Spangled, Chamois and White Spangled.

Weights

Cock	5.5 pounds (2.50 kg.)
Hen	4.4 pounds (2.00 kg.)

Trio Vorwerkhuhners

Vorwerkhuhner

The Vorwerk breed, as they are also known, was developed by Oskar Vorwerk, hence the name. In 1900, he began experimenting with mixtures of Lakenvelders, Buff Sussex, Buff Orpingtons and Blue Andalusians. He was aiming for a dual purpose, large bodied bird with good egg producing qualities. The Vorwerkhuhner has a Buff Lakenvelder color pattern.

Weights

Cock	6.6 pounds (3.00 kg.)
Hen	5.5 pounds (2.50 kg.)

Westfalische Totleger Pair
by
Mary Ellen Johnson

Westfalische Totleger

This breed was originally called the Every Day Layer. The Westsfalische Totleger was found in the Ravensberg region in early 19th century and is related to the Brakel of Belgium. Westfalische Totlegers are high egg producers of white eggs. They have distinctive Rose Combs and are weather tolerant. This bird is bred in Silvers and Goldens.

Weights

Cock	5.5 pounds (2.5 kg.)
Hen	4.4 pounds (2.0 kg.)

White Emdens

Emdens were bred for many centuries in western Europe. They may have come from Italy and are possibly related to Crested Romans. White Emdens were developed as sports from White Fronted and White Breasted geese. Because of their commercial qualities, they grew in popularity and were known as Geese of Emden.

White Emdens were recognized in the *American Standard of Excellence* in 1874.

Weights

Gander	30-34 pounds (13.60 - 15.50 kg.)
Goose	20-22 pounds (9.10 - 10.00 kg.)

Pair White Emden Geese

Barnvelders

Barnvelders were developed around 1850 by crossing a local breed with the Malay, Brahmas and Langshans. The original breeding took place in Barnveld village, Veluwe county in the province of Gelderland. This is in the heart of the poultry industry and near the University of Poultry Husbandry. These chickens are known as a dual purpose fowl with good meat quality and excellent egg laying, even in the winter. They were imported into England in 1921 and are recognized in Dark, Black, White and Blue Laced. Recognized in the *American Standard of Perfection.*

Weights

Cock	7 pounds (3.20 kg.)
Hen	6 pounds (2.70 kg.)

Barnvelder Cock
Barnvelder Hen

Dutch Booted Bantam

Trio Mille Fleurs

This is another controversial chicken because Belgium claims them as their Belgian Bearded d'Uccle and Mille Fleurs. The Mille Fleurs were imported into the United States from Germany. In the Stromberg catalog, they were featured as a French breed, the name Mille Fleur meaning, *Thousand Flowers.* The Dutch claimed these Bantams on the basis that they had been known in Holland for five centuries.

White Dutch Booted Cock

Dutch Booted arrived in Europe in their original Mille Fleur color. There is also a belief that the Chabo Bantams of Japan share an ancestry with these Bantams. A characteristic of this bird is its vulturine hocks which distinguishes them from other Bantams. The Dutch Booted is bred in Mille Fleur, Lemon Mille Fleur, Blue Mille Fleur, Mottled, Blue Mottled, Red Mottled, Yellow Mottled, Black, White, Self Blue, Cuckoo, Pearl Grey, Partridge, Silver Duckwing and Gold Duckwing.

The Mille Fleur colored Dutch Booted Bantam is found in German, Danish and Swiss breeding Standard books. They were recognized as Mille Fleur in the *American Standard of Perfection* in 1914.

Weights

Cock	850 grams
Cockerel	750 grams
Hen	650 grams
Pullet	550 grams

Dutch Owl Beards

The Dutch Owl Beards are an extremely rare breed but are becoming a bit more popular.There seems to be a relationship with the Polish. A point of distinction is their specific V comb and their beard. They are recognized in Black, White, Blue Laced, Cuckoo, Gold Speckled, Silver Speckled, White Laced Buff, Golden Pencill and Silver Pencill.

Weights

Cock	5.5 pounds (2.5 kg.)
Hen	4.0 pounds (1.8 kg.)

Frisian

The Frisian is one of the oldest breeds in Holland and was developed in the province of Friesland. They are closely related to the Drente Hoen. For many years they were only one color but now are recognized in Pencilled, Golden Pencilled, Silver Pencilled, Red Pencilled, Yellow-White Pencilled, Black, White, Cuckoo and Mottled.

Weights

Cock	3.5 pounds (1.6 kg.)
Hen	2.8 pounds (1.3 kg.)

Top
Cuckoo Dutch Owl Beard Cock

Middle
Blue Dutch Owl Beard Hen

Bottom
Silver Pencilled Frisian Hen

Hamburgs

Golden Pencilled Hamburg Hen

In Holland, the Hamburgs are known as the Holland Hoens. Aldrovandi described them as Turkish Poultry (Gallus Turcicus). They are a very old breed and identified as a specific breed in an early English painting. During the 18th century in Lancashire and Yorkshire, England, the Hamburgs were known as Pheasants and Mooneys.

References to Bolton Greys and Bolton Bays were possibly describing the earlier versions of Pencilled Hamburgs. They were also known as Chittaprats, Creoles, Dutch Every Day Layers and Pencilled Dutch Fowl. Books published in 1853 and 1857 have pictures of Hamburgs with crests called Crested Hamburgs. There may be some relationship with the Polish, before their present type and color was established.

The Hamburgs may have been derived from Paduan, a mixture of Sumatra, Campines and Totleger. They were developed in Holland with the Latin name of Gallus Pativanus or Bativanus, which was the name of a Netherlands tribe during Roman occupation.

These chickens are excellent white egg layers for most of the year with the exception of winter and are bred only with Rose Combs. Hamburgs are recognized in Holland as Gold Spangled, Silver Spangled, Gold Pencilled, Yellow Pencilled, Yellow White Pencilled, Blue Gold Pencilled, Black, White, Blue and Cuckoo.

Golden Pencilled Hamburg Cock

Weights

Cock	5 pounds (2.25 kg.)
Cockerel	4 pounds (1.80 kg.)
Hen	4 pounds (1.80 kg.)
Pullet	3 1/2 pounds (1.60 kg.)

Above
Kraaikoppen Cock

Below
Kraaikoppen Hen

Kraaikoppen

The Kraaikoppen is an old Dutch rare breed (A.D. 1400) developed from crested poultry. They are depicted in old paintings by Jan Steen in 1650. Developed in the province of Gelderland, the Kraaikoppen is related to the Twente Hoen and other Dutch breeds.

In other countries they are called the Bred Fowl, named after that city in Holland. They are a good breed to caponize and though considered a combless breed, it does have a small Rose Comb.

The Kraaikoppen has wide nostrils, feathered feet and vulture hocks. They are bred in Black, Blue Laced, White and Cuckoo.

Weights

Cock	6 1/2 pounds (2.95 kg.)
Hen	5 1/2 pounds (2.50 kg.)

Mating of Golden Lakenvelders

Lakenvelders

The ancestry of the Lakenvelders is controversial with reports from Holland and Germany claiming their origin. In Holland, it is thought of as a Dutch breed developed in Shakespeare of Laneveld in the province of Utrecht near the year 1700. Germany claims their heritage came from Westphalia about 1835. I have found nothing specific regarding the breeds involved to produce the Lakenvelders.

They are bred in Black, White, Golden and Blue and were first shown in England in 1902. In the United States, Lakenvelders were admitted to the *American Standard of Perfection* in 1939.

Weights

Cock	5 pounds (2.25 kg.)
Cockerel	4 pounds (1.80 kg.)
Hen	4 pounds (1.80 kg.)
Pullet	3 1/2 pounds (1.60 kg.)

Trio Frizzled Krulbaardkuifhoen

Netherlands Krulbaardkuifhoen

To many people the re-creation of a breed is beyond comprehension. It requires knowledge of genetics and tremendous patience to make it happen. The development of this Bantam took A.J. Boland, Netherlands, over ten years to accomplish. The Netherlands Krulbaardkuifhoen is an uniquely beautiful Bantam. This re-creation is a great credit to Boland that will arouse the desire of many poultry fanciers to own them. When I visited him in 1989, he had none to sell but hopefully, his stock will increase.

Gold Partridge Old Dutch Cock

Old Dutch Bantams

The Old Dutch is known as the first Bantam of Holland. The exact history isn't known but they were developed before the year 1800. The original color was Bankiva, the color of Jungle Fowl. They may have been one of the first domesticated chickens to branch off from Jungle Fowl. I know the first miniature chickens were named after the port city of Bantam, Java. Dutch ships brought spices to Holland and perhaps may have brought the Bantams that are known today as Old Dutch.

Old Dutch Bantams are one of the most popular breeds exhibited in Holland and are bred in Partridge, Silver Partridge, Blue Partridge, Blue Silver Partridge, Cuckoo Partridge, Red Shoulder Blue Silver Partridge, Red Shoulder Blue Silver Partridge, Red Shouldered White, Salmon, White, Black, Self Blue and Cuckoo.

Silver Duckwing Old Dutch Cock

Weights

Cock	550 grams
Hen	450 grams

Polish

Holland has always been given credit for the development of Polish chickens. The name may have been derived from the Po River in Italy. The country of Poland had nothing to do with their development. According to *The Poultry Book*, they were developed in Padua, Italy. This would explain Aldrovandi's name of Paduan Fowl.

White Polish Cock

The Polish are thought to have descended from the Gallus Giganteur of Sumatra with a claim that they are of Spanish extraction, and the Spanish introduced them in Holland. These chickens were found in the county of Kennerland in north Holland in 1475 and

have been known in Holland for centuries. There is no doubt the Dutch perfected them since they don't resemble those fowl pictured by Aldrovandi.

In Holland, the Non-Bearded and Bearded Polish are regarded as two specific breeds. They are considered good layers of white eggs and bred in White Crested Black, White Crested Blue, White Crested Cuckoo, White Crested White, White Crested Mottled, White Crested Blue Laced, all Blue Laced and all Cuckoo, and Black Crested Whites, which are very rare.

White Polish Hen

The first Polish were recognized by the *American Standard of Excellence* in 1874. In 1883, the following varieties joined them, Bearded Goldens, Bearded Silvers, Bearded Whites, Bearded Buff Laced and in 1938, Buff Laced Non-Bearded.

Weights

Cock	6 pounds (2.7 kg.)
Cockerel	5 pounds (2.3 kg.)
Hen	5 pounds (2.25 kg.)
Pullet	4 pounds (1.80 kg.)

Very rare White Crested Cuckoo Polish

Twente Greys

The Twente Greys was developed during 1850 to 1860 in Enschede, in the province of Overijssel. The original breeding is from Malyas and Silver Partridge Leghorns. In Bensheim, Germany, they are known as Bieshuhners.

The Twente Greys is a popular exhibition breed and considered a very good egg layer with more than 200 eggs a year.

Weights

Cock	5.5 pounds (2.5 kg.)
Cockerel	4.4 pounds (2.0 kg.)

Above
Twente Grey Cock

Below
Twente Grey Hen

Pair Dutch Welsumers

Welsumers

The Welsumers came after the Barnvelders. They originated in the area of Welsum near the river Ijssel in Overijssel and are a blend of Barnvelder, Rhode Island Reds, Partridge Cochins, Partridge Wyandottes and Partridge Leghorns. The first Welsumers came to England in 1928. They are considered layers of nice, large, brown eggs. Recognized in the *American Standard of Perfection* in 1991.

Weights

Cock	7 pounds (3.20 kg.)
Hen	6 pounds (2.70 kg.)

Hungary

Magyar Chicken

Magyar means *Hungarian* and these native chickens have prevailed in Hungary for over 100 years. Sir Edward Brown related them to the Italians or the chicken we know of as Leghorns. They are bred in White, Light Brown and Cuckoo and considered fair layers.

Weights

Cock	6 1/2 pounds (3.0 kg.)
Hen	5 1/2 pounds (2. 5 kg.)

Magyar Geese

Magyar geese were known in the time of the Romans. Their egg production was small, 25 to 30 eggs, but with advanced breeding techniques, the production has increased to 50 eggs a year. These geese are used as the male line in different Hungarian hybrids for efficiency to produce large livers, meat and feathers.

Top
Cuckoo Magyar Pair

Middle
Light Brown Magyar Pair

Bottom
Magyar Geese

Weights

Ganders	15 pounds (7.0 kg.)
Goose	12 pounds (5.5 kg.)

India

Aseels

Aseel is a Hindu word meaning *high born or aristocratic.* The Mohammeds called this chicken Aseel Mooruh or *exceeding noble cock*. They were probably the first fighting stock and may be related to Madagascar Game Fowl, a vicious fighting breed.

In the early days, English fanciers referred to Aseels as true Indian Game Fowl. Darwin's claim that modern chickens originated from Jungle Fowl is logical and most fanciers agree the theory applies to most breeds. However, for fowl such as Aseels, Malays, Cochins, Brahmas and Langshans, it is questionable. In 1926, Thomas McGree wrote in *The Poultry Book*, the theory the Aseels was the first fighting chicken. The Institutes of Mann (1000 B.C.) claims the Aseels were the first domesticated chicken.

There are several strains of Aseels in differrent areas of India and all vary in size and color. In 1981, Aseels were accepted into the *American Standard of Perfection* in Black Breasted Red, Dark, White and Spangled.

Weights

Cock	6 pounds (2.70 kg.)
Cockerel	5 pounds (2.25 kg.)
Hen	5 pounds (2.25 kg.)
Pullet	4 pounds (1.80 kg.)

Brahmas

The Light Brahmas are well known around the world but there are also Dark Brahmas and Buff. Like Cochins that came from China and India, Brahmas have changed considerably since 1846, when they had only a few feathers on their shanks. The first Brahmas came to England in 1853 and originated from the Indian side of the Brahma Pootra River, opposite Chittagong, China.

The first *Brahmapootras* were imported in 1846 from Lukpoor, a good distance from Brahma Pootra River. Dr. John C. Bennet was the first to exhibit them on October 24, 1850, at Fitchburg Railroad Depot Hall.

Light and Dark Brahmas were recognized in the *American Standard of Excellence* in 1874. The Buff was developed later and acknowledged in the *American Standard of Perfection* in 1924.

Weights

Cock	12 pounds (5.45 kg.)
Cockerel	10 pounds (4.55 kg.)
Hen	9 pounds (4.10 kg.)
Pullet	8 pounds (3.60 kg.)

Opposite Top
Black Breasted
Red Aseel Cock

Opposite Bottom
Spangled Aseel Cock

Right Top
Dark Brahma Cock
Light Brahma Hen

Green Jungle Fowl Cock

Jungle Fowl

While Jungle Fowl are never exhibited in competition, according to Darwin's theory, they are the starting point of all chickens. Jungle Fowl is a small breed and has different species.

Red Jungle Fowl	*Gallus gallus*
Grey Jungle Fowl	*Gallus sonneratii*
Green Jungle Fowl	*Gallus varius*
Ceylonese Jungle Fowl	*Gallus lafayettii*

Kadakanath

The Kadakanath is a newly developed breed in India. Without specifics, I can only present these pictures.

Kadakanath Cock
Kadakanath Hen

Iran

Rumpless Fowl

This unique fowl was known to Aldrovandi in the year 1600. He called them Persian Fowl implying Persia was their land of origin. In Tegetmeier's book, 1867, he quotes the writer Temminck as stating these Rumpless Fowl were found in the forests on the Island of Ceylon and known as Wallikikilli or Cock of the Woods. This fowl was known on the Isle of Mann as well.

Bement, in his 1856, *The Poulterer's Companion*, refers to them as native of Persia, *(now Iran)*. In Belgium, they are known as Walloons *(Hedge Fowls)*, and another name applied to them was Manx Rumples.

I have pictures of Rumpless Fowl from Japan, France, Germany and Holland but they are identified by different names. This chicken is bred in a variety of colors and is not recognized in any breeding standard. The Rumpless Fowl is a fair layer of tinted to brown eggs, and the following weights are by a breeder who raises them.

Weights

Male	4 1/2 pounds (2.00 kg.)
Female	3 1/2 pounds (1.60 kg.)

Top to Bottom
Uzurao Black Rumpless Pair
Iran Lari Fowl
Umumi Fowl

Anconas

In the book, *Domestic Races of Poultry* by Sir Edward Brown, he states the Anconas is a variety from the common Italian fowl, likely the result of many crossings. Lewis Wright, in early writings, thought Anconas could be sports from Black and White Minorcas. It is most often reported that they came from Marches, a port city of Ancona, Italy. They were imported into England in 1848 and exhibited in 1851. The Anconas were imported into the United States in 1888 and South Africa in 1902.

Anconas enjoyed great popularity in the 1920's. In his book, *History of Anconas*, Roy Van Hosens refers to the *Anconas World* magazine which has been in existence since 1910, and to the Anconas Club. Cecil Shepard, Berea, Ohio, was a member of that club and imported his first stock in 1906. He soon became the leading breeder of that era. In 1920, he wrote *'In One Year an Income of $4223.00 on a Town Lot'.* That was a very good income in those days. In 1922, Shepard bought a small farm and in one year raised 18,000 Anconas and reported an income of $10,000.

In the 1930's, Shephard's Anconas continued their popularity but eventually, White Leghorns won in nation wide popularity. The Single Comb Anconas was admitted to the *American Standard of Perfection* in 1898 and the Rose Comb in 1914.

Top
Anconas Cock

Below
Anconas Hen

Weights

Cock	6 pounds (2.70 kg.)
Cockerel	5 1/2 pounds (2.50 kg.)
Hen	5 pounds (2.25 kg.)
Pullet	4 1/2 pounds (2.00 kg.)

Crested Roman Geese

Crested Roman Gander

Sir Edward Brown thought Crested Roman Geese were a very early development from the Greylag. Early writers such as Homer, Livy, and Lucretius state that these geese were kept in Rome and considered sacred to Juno which meant the Priests used them for food.

In England, the Roman Geese were very popular and perhaps they were the heritage of other varieties, but this has not been proven. Reference has been made to their association with early development of Sebastopol geese but when the crests or tufts appeared is not clear.

These geese are recognized in the *British Poultry Standard,* but without the crest. In the United States the Crested Roman Geese were accepted in the *American Standard of Perfection* in 1977.

Weights

Adult gander	12 pounds (5.45 kg.)
Young gander	10 pounds (4.55 kg.)
Adult goose	10 pounds (4.55 kg.)
Young goose	9 pounds (4.10 kg.)

Leghorns

Leghorn Cock
Leghorn Hen

Sir Edward Brown reported the first White Leghorns came to the United States in 1828 or 1829. N.P. Ward, New York , imported the first Brown Leghorns in 1835. In *The Book of Poultry*, 1926, it was reported that these chickens were observed in 1815 in diverse colors and shapes, along the Mediterranean coast. Black Leghorns appeared in the U.S. in 1872 and the Buff in Denmark in 1883. The English imported their first Leghorns from the United States in 1870.

The British began breeding work with Buff Leghorns and various crosses of the Buff including Buff Cochins. Silver Leghorns were developed by crossing the Yokohamas, Silver Grey Dorkings and Silver Duck Wing Games.

Spangled Leghorn Cock

Exchequor Leghorns were developed in Scotland from sports of White Leghorns. There is a contradictory report that Exchequor Leghorns was developed with a cross that involved the Anconas.

Blue Leghorns were exhibited under the name Italians in a London poultry show in 1908 and at the first U.S. poultry show held in November, 1949. The Dominque, Mottled, Partridge and Pyle Leghorns are extinct.

All Leghorns are considered good layers but White emerges as number one in the world. In the United States, efforts to improve egg production resulted in breeders trapnesting them to select the very best producers. Many commercial hybrids evolved from this chicken.

In 1912, Professor James Dryden, Oregon State University, perfected the first Leghorn to lay 300 eggs in one year. J.A. Hanson, Oregon, developed layers with records of over 300 eggs a year and a trapnest era designated ROP (Record of Performance) established the popularity of White Leghorns as the leading commercial layer. The majority of the world's eggs are produced by the hybrid and strain crosses of this popular variety.

The *American Standard of Excellence* of 1874 shows that Single Comb White, Brown and Black Leghorns were recognized and in 1883, Rose Comb Brown Leghorns joined them. The Rose Comb White Leghorns were identified as a new breed in 1886 and in 1894, Single Comb Buff and Single Comb Silver Leghorns were recognized.

In 1929, the American Poultry Association acknowledged Single Comb Red, Single Comb Red Black Tailed and Single Comb Colombian Leghorns.

Weights

Cock	7 1/2 pounds (3.40 kg.)
Cockerel	6 1/2 pounds (2.95 kg.)
Hen	5 1/2 pounds (2.50 kg.)
Pullet	4 1/2 pounds (2.00 kg.)

Japan

Chabo Bantams

I owe full credit to Anna Banning Vogelpoel, Holland, for the history of Chabo Bantams. She is the author of the book *Japanese Bantams*. In some countries, including the U.S., the Chabo's are known as Japanese Bantams. The first stock arrived from the China coast from 1603 to 1636. From 1636 to 1867, Japanese ports were forbidden to the outside world. However, the Dutch had communication with the island Decima from 1641 to 1867. Because of this Holland was the first to get the Chabo Bantams.

Black Tailed Japanese Cock
Black Tailed Japanese Hen

In ancient times these beautiful and rare Bantams were excellent *honoring gifts* and artist Jan Steen included the Buff Black Tailed Chabo in a 1600 painting, confirming the early arrival of these birds in Holland.

The Japanese breed them in many colors including Frizzle and Silky feathered species. In the United States, they were accepted in the

American Standard of Excellence in 1884 in Black, White, and Black Tailed. Respectively, the following were accepted, Greys in 1914, Mottles in 1947, Black Tailed Buffs in 1982.

Weights

Cock	18-22 oz.(510 - 620 grams)
Hen	14-18 oz. (400 - 510 grams)

Red Frizzle Trio

Frizzles

Nothing is known about the origin of Frizzles. An 1856 book, The *American Poulterer's Companion* credits Japan as the land of origin but they were also known in Java, Sumatra, Philippines, India and southern Asia.

Frizzles are an old breed and have been referred to by

early writers as Frisland, Crisp Feathered Fowl, Caffie and Frizzled Fowl. They are bred in Black, Red, Buff and White and were recognized in the *American Standard of Excellence* in 1874.

Black Frizzle Hen

Weights

Cock	8 1/2 pounds (3.85 kg.)
Cockerel	7 1/2 pounds (3.40 kg.)
Hen	6 1/2 pounds (2.95 kg.)
Pullet	5 1/2 pounds (2.50 kg.)

Phoenix

Though this birds heritage is Japan, the Japanese don't have a breed by that name. They are distinguished by their Single Comb, in contrast to the Walnut Comb of the Yokohamas. In Japan they are known as Minohikis. The Phoenix was imported into Germany in 1878.

It is my conclusion that the English and Germans deserve the credit for the current type. They do not molt their tail covert feathers nor their saddle feathers. Some bloodlines of the Phoenix don't carry this gene and do molt these feathers. I believe the Phoenix is more closely related to the Onagadori than the Yokohmas. They can develop tail feathers three to four feet long. The Silvers were accepted in the *American Standard of Perfection* in 1965 and Golden Phoenix was recognized in 1983.

Silver Phoenix Cock

Weights

Cock	5 1/2 pounds (2.50 kg.)
Cockerel	4 1/2 pounds (2.00 kg.)
Hen	4 pounds (1.80 kg.)
Pullet	3 1/2 pounds (1.60 kg.)

Black Breasted Shamo Cock

Shamos

Shamos are one of the oldest game fowls known. The basis of origin is the Malays which was imported into Japan from Siam for cockfighting. They were line bred for over 500 years. There are six types of Shamos in Japan of varying size and color. Shamos were imported into the United States in 1874 and welcomed by cockfighting enthusiasts.

This is a rugged fowl, very tall, standing 30 inches high. They are recognized for exhibition and were admitted to the *American Standard of Perfection* in 1981.

Weights

Cock	11 pounds (5.00 kg.)
Cockerel	9 pounds (4.10 kg.)
Hen	7 pounds (3.20 kg.)
Pullet	6 pounds (2.50 kg.)

Tosa Onagadori

The history of the Onagadori is very interesting. It was first known as the Nag-O-Dori which means *long tail fowl.* This is the most unusual of all Japanese fowl because of their unique feathering characteristics and beautiful long tails. While the Onagadori is a poultry treasure of Japan, there are questions about its early origin. Fowls with five foot tails were mentioned in old Chinese manuscripts so breeders persisted and crossed the Onagas with the Skogoku which produced a red feathered long tail fowl call the Aka-Zasa.

Credit for the development of Onagadori is given to Reiman Takechi who lived in 1655 in the Tosa province during the Feudal period three hundred years ago. In this time, it was customary to hold royal pageants for the Shogun who would demand long chicken feathers to pay him tribute. These were used to adorn his ceremonial spears.

It was Reiman's desire to please his Lord so he experimented with the Jedori, a common fowl and produced birds with tails several feet in length. He was overjoyed with his success but hopes of glory vanished when a civil disturbance changed the situation and the province became very poor. Food was scarce but Reiman was able to beg feed for his flock. A monument was erected to pay tribute to Reiman Takechi for saving the Onagadori. After Feudalism, fanciers continued breeding programs and found the genetic combination that led to extremely long tails.

Very well displayed Tosa Onagadori Cock

In 1908, the Long Tailed Fowl Preservation Association was formed and on March 7, 1923, Onagadori was honored as the National Treasure Bird by the Ministry of Education. During the Showa reign, the Japanese government declared Onagadori a rare bird and established a museum in its honor. The name Naga-O-Dori was changed to Tosa Onagadori when it was commemorated.

There are many breeds of long tail fowl in Japan; Tosa Onagadori, Shokoku, Totenko, Tomaru, Kashiwa, Ohiki, Minohiki, Satsuma Dori Kurekodori and the Shamokan. Only the Tosa Onagadori is capable of growing a long, permanent tail with ten to fifteen feet in length being common. There are reports of older males with tails of 25 to 30 feet.

The adult male is Single Comb, has white ear lobes, long bushy hackles and a very long tail. The hackle feathers cover the shoulders and reach the base of the tail. The male sheds its body feathers at times, but never the tail feathers which grow yearly. A bird has 26 tail feathers, however, a male with 20 feathers is still considered a good bird for judging.

10 month old Onagadori

The breeding of the Onagadori is done using a spare cock whose tail is cut because a long tail hampers mating. Natural breeding is always used, never artificial insemination. The female grows a normal tail and molts completely every year. It is she that carries the long tail gene, transmitting its characteristics to the chicks. Onagadori are managed in very small groups.

The care, housing, feeding and showing of this bird is a special challenge and the Japanese dominate this area. The male bird of five to six months of age is segregated by placing it in a rearing cage called a Tomebako, meaning *clear box*. This is a tall wooden box with swinging glass doors. A large wooden dowel serves as a perch and is placed near the top of the Tomebako. The box is suspended above the ground and food and water are placed in the front. A removable tray is included for the droppings. Usually, the box is 180 centimeters high, 80 centimeters wide and only 20 centimeters deep. These quarters make it easy to observe the beauty of the bird but the bird is really a captive. They are taken from the box for airing and exercise or for exhibition. When removed, great care is taken to protect their tails which are often coiled, suspended with string and carried by an assistant.

Caging of this manner does encourage growth of tail and saddle feathers. The sickle feathers grow at a rate of three feet per year and the saddle, at one foot per year. A bird raised on the ground could never attain maximum growth. My friend, Mr. Kawamua of Japan, feeds a unique mixture of live mud fish and boiled pork to his Tosa Onagadori and exhibited them at the 1988 World Poultry congress held in Nagoya, Japan.

No breeder outside of Japan has been successful in developing 15 to 30 foot tails on the Onagadori. It's been reported that John Kriner, Pennsylvania, did raise one with a tail of 12 feet. Poultry fanciers will have to depend on the skillful Japanese fanciers to carry on the tradition of perpetuating Tosa Onagadori which are bred in Silver Duckwing, Shirozasas, Akazasa, Black Red or Red Breasted, Shyo Shu, Akai Red, Shiro and White.

Weights

Cock	4 pounds (1.8 kg.)
Hen	3 pounds (1.3 kg.)

Darker Red Shouldered Yokohama Cock

Yokohamas

The Yokohamas heritage is Japan but the Japanese don't have a chicken by this name. This bird shares an ancestry with the Phoenix and derived its name from the city of Yokohama, Japan. They were also known in Korea in the year 1000.

Yokohamas were imported into France in 1864, Germany in 1868 and England in 1884. A 1903 book reports that the Germans bred them to great perfection. It indicates crossings by both the Germans and English utilizing Polish, Hamburgs, Minorcas, Old England and long tailed Langshans, and shows a White and Multi-Colored Yokohama. These birds can attain tail feathers of 40 inches in length.

The *British Poultry Standard* recognized the Yokohama as a Single Comb breed and not the Phoenix. Yokohamas are distinguished by a Walnut Comb, small and set forward. They were accepted in the *American Standard of Perfection* in 1981 in White and Red Shouldered.

Weights

Cock	4 1/2 pounds (2.00 kg.)
Cockerel	4 pounds (1.80 kg.)
Hen	3 1/2 pounds (1.60 kg.)
Pullet	3 pounds (1.35 kg.)

Java

First Bantams

The origin for the name of our miniature or dwarf fowl called Bantams, came from the city of Bantam, Java. In 1595, Dutch ships brought spices and the first chickens, called Bantams, from the Island of Java. It was then that the earliest Bantams were introduced in Holland. Harrison Weir, England, wrote that early Bantams in the United States were known as Crow Chickens, and the miniature fowl resembled Old English Games. Bement, in 1856, wrote that some of the first known Bantams resembled the Bankazvia, commonly called Jungle Fowl.

Many of the early English Bantams were feather-shanked. Marco Polo noticed the small fowl known as Silkies in the 13th century. Aldrovandi reported on dwarf chickens as did Moubray in the early 1800's. Sir John Seabright, England, developed the Sebright Bantam and is credited with the popularity of these birds.

Black Java Cock
by
Mary Ellen Johnson

Java Chicken

Professor John Skinner reporting in *American Poultry History* said Java were imported into the United States from Java, East Indies, in 1835. However, ancient poultry books have made the Java a chicken of controversy. I've found it mentioned that the Java is the same fowl as Cochins. Does this mean they had feathered shanks? Adding to the controversy, it's said the Barred Plymouth Rocks was used in its development. This is questionable since that breed was not perfected until a later date. I also found that the Java was modified before it was recognized by the American Poultry Association as a new breed.

Mottled Java Cock
by
Janice Brent

The Poultry Book, 1912, tells the story of a Missouri doctor who delighted in owning fine black fowl, called Java. The doctor would not sell any of the progeny nor permit any to grace the yards of his neighbors. A family came into possession of three eggs that were borrowed by their coachman from the doctor. This gave them their start. In 1857, the family moved from Missouri, and took their Java to New York. From then on, the Java was bred in large numbers.

J.Y. Bickell, New York, reported that the Mottled Java was an accident. The crossing that produced them was a white hen sired by a Black Java cock. Both the Black and Mottled Java were accepted as new breeds in the *American Standard of Excellence* in 1884.

Weights

Cock	9 1/2 pounds (4.30 kg.)
Cockerel	8 pounds (3.60 kg.)
Hen	7 1/2 pounds (3.40 kg.)
Pullet	6 1/2 pounds (2.95 kg.)

Malaysia

Indian Runners

The Indian Runners were first known as the Penguin duck in India. It is a fitting name because their tail stance is reminiscent of a Penguin. They were introduced to England when a ship brought them from Malaysia, but conflicting information says they came to England from Calcutta and Bombay, India. Unfortunately, there is no history of their development.

The first to be exhibited was the Fawn Indian Runners in 1876 at the Dumfries Poultry Show. In 1907, the Indian Runners Club was organized in England.

These birds are excellent layers and are known as the Leghorn of the duck family. Fawn and White Runners were recognized by the American Poultry Association in 1898. Pencilled and White Indian Runners were accepted in the *American Standard of Perfection* in 1914.

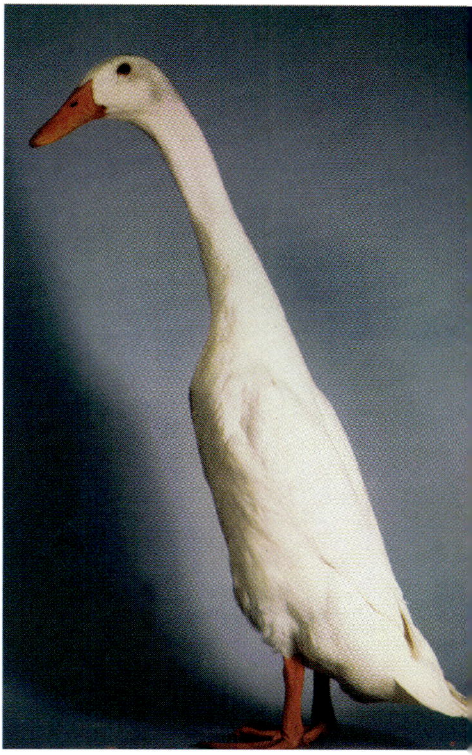

Weights

Adult drake	4 1/2 pounds (200 kg.)
Young drake	4 pounds (1.80 kg.)
Adult duck	4 pounds (1.80 kg.)
Young duck	3 1/2 pounds (1.60 kg.)

Above
White Indian Runner Hen

Below
Fawn and White Indian Runners

Above
Black Breasted Red Malay

Below
Weaten Malay

Malay

The Malay, without a doubt, was the early foundation stock of fighting chickens. They can stand as tall as 30 inches. Sir Edward Brown wrote that the Malay was from the Malay Peninsula. Other writers agree, though Doyle, places them in India and the Philippines as well as Malay, from which came their name. Bement thought the Malay originated from Kulm or Gigantic Fowl, and that they may be the heritage stock for Chittagongs, Cochins, St. Jago and Brahmas.

Tegetmeir, a famous poultry writer, thinks the origin of the Malay came from the wild species of Gallus Giganteis of Temminick, while Moubray considers them a wild fowl and refers to them as Chittagongs.

Prior to the arrival of Cochins in England, the Malay was recognized as the largest fowl and enjoyed a great deal of popularity. Colonel Cykes imported the first Malay into England in 1831. They were shown at the first English poultry show in 1845 and recognized in the *British Standard* in 1865.

In the United States, the Black Breasted Red Malay was recognized in the *American Standard of Excellence* in 1884 and Spangled, Black, White and Red Pyles were accepted into the *American Standard of Perfection* in 1981.

Weights

Cock	9 pounds (4.10 kg.)
Cockerel	7 pounds (3.20 kg.)
Hen	7 pounds (3.20 kg.)
Pullet	5 pounds (2.25 kg.)

Mexico

Turkeys

Turkeys, though known in areas of South America, are known to have had their start in Mexico. It seems logical that the wild turkey in the United States came from Mexico. Rene Randonniere wrote that the first turkeys were observed in North America in 1564 and Christopher Columbus found them on the island of Guanaja in 1502.

Sir Edward Brown reported that in Venezuela,1499, the first turkeys were obtained by Pedro Nino who purchased them from the Indians for four glass beads per fowl.

Turkeys found their way to Spain in 1500 and Cortez, who went to Mexico in 1520, saw an abundance of this fowl. They were introduced into Cuba in 1511, Yucatan in 1517, England in 1524, Italy in 1557 and France in 1570. Bartolomeo Scappi, a chef for Pope Pius V, had several turkey recipes which increased their popularity. King Charles IX of France began breeding them in the forest of St. Germain.

The name *turkey* was given to this breed in Spain. *Turkish* meant foreign in the early days and merchants trading abroad were known as turkey merchants, hence the name. Contrary to the story that Christopher Columbus brought the first turkeys to America, they were already in the United States.

Above
Bronze Turkeys

Middle
White Holland Turkeys

Bottom
Black Turkeys

The Cambridge Bronze was developed from the stock of turkeys that reached Europe and arrived in England in 1524. In the U.S., the American Bronze evolved from wild turkeys. White turkeys appeared in France,

Royal Palm Turkeys
Toms

Austria, Hungary and Holland. The United States received white turkeys from Holland, thereby the name, White Hollands. Black turkeys are credited to Mexico. The Narragansetts were developed in Rhode Island but there is no history of their origin.

The Bourbon Reds were developed by J.F. Bargee, Bourbon county, Kentucky, from a cross of the Bronze and Buff turkey. They were called Bourbon Butternuts but the name Bourbon Red finally prevailed.

Royal Palms were developed in Florida from a single poult, a sport, originating from a flock of assorted turkeys. The Beltsville Whites were developed by the United States Department of Agriculture, Beltsville Research Station. The following turkeys were recognized in the *American Standard of Excellence* in 1874, Bronze, Narragansetts, White Hollands, Blacks and Slates. Bourbon Reds were accepted in the 1909 *American Standard of Perfection* along with Beltsville Whites in 1951 and Royal Palms in 1977.

Weights

Bronze

Adult tom	36 pounds
Young tom	5 pounds
Adult hen	20 pounds
Young hen	16 pounds

Narrangansett, White Holland, Black, Slate

Adult tom	33 pounds
Young tom	23 pounds
Adult hen	18 pounds
Young hen	14 pounds

Royal Palm

Adult tom	22 pounds
Young tom	16 pounds
Adult hen	12 pounds
Young hen	10 pounds

Bourbon Red Turkeys

Ocellated Turkey

The Ocellated turkey is regarded as a wild turkey from Belize, Honduras, Guatemala and southern Mexico. It is claimed they were first discovered in 1920 and appeared in the U.S. in 1930. The San Diego Zoo was the first to raise them in captivity.

In the United States, ownership of an Ocellated turkey is limited to a few zoos and a dozen dedicated aviculturists because they are an endangered species. The Ocellated turkey is more colorful than the wild turkey.

Top
Ocellated Turkey Tom

Middle
Slate Turkeys

Bottom
Wild Turkeys

Norway

Jaer Fowl

Development of the Jaer began in the early 1920's at the Jaer Breeding Station, from which they took their name. The objective was to produce a smaller, more efficient egg layer. In a five year test period, they averaged 190 eggs a year. The Jaer Fowl is bred in limited numbers and the commercial egg laying strains dominate the production of eggs in Norway. They were also developed as an auto sexing breed which means sex can be determined by visual observation.

Weights

Cock	5 pounds (2.25 kg.)
Hen	4 pounds (1.80 kg.)

Jaer Fowl Pair

Paraguay

Langosta

The Langosta chicken was developed by Emil Bauknecht of Paraguay about 1965. The objective was to breed a meat type chicken with good flesh qualities. Successful marketing started in 1973. After the death of the leading Langosta breeder in 1976, the poultry farm was neglected and by 1984 the Langosta fowl was nearly forgotten. New ownership of the farm increased production and now the Langosta supplies ten percent of the market in Paraguay. Egg production however, is 140 eggs per year.

Langosta Cock

Poland

Greenleg Partridge

In Sir Edward Brown's book, there is mention of the Polish Greenfoot breed. He refers to their introduction from Asia with crosses to Mediterranean races. The Greenleg Partridge appears to be related to the native chicken of Finland and is considered a fairly good layer.

Weights

Cock	5 1/2 pounds (2.25 kg.)
Hen	4 1/2 pounds (2.00 kg.)

Greenleg Partridge Cock
Greenleg Partridge Hen

Polbar

The only reference I can make to the Polbar is a picture supplied by the Polish Gene Bank for the preservation of an endangered species. These chickens are rated as fair layers.

Weights

Cock	6 pounds (2.70 kg.)
Hen	5 pounds (2.25 kg.)

Polbar Cock
Polbar Hen

Romania

Sebastopol Geese

When considering the origin of a particular breed of chicken or waterfowl, ancient history comes into play. The Danubians invaded Greece 1300 B.C. and introduced white geese to that region. I think these might be associated with Crested Roman geese that originated in Italy thus it is logical to consider these as ancestors of the Sebastopol.

Sebastopol Gander

Sir Edward Brown reports British breeders to have imported Sebastopol geese at the time of the Crimean War, 1859 to 1860. He places their origin in the area of the Danube River and feels they prevailed in countries bordering the Black Sea. The name Sebastopol is associated with a port by that name on an island in the Black Sea which means they could be of Russian origin.

I have credited Romania as their homeland because that is the only nation to claim them. Though exhibited in Europe, the Sebastopol is not popular. I believe there are more Sebastopol geese in the United States than in any other nation.

For a period of time they were know as Danubian and Ripple geese. John Kriner, Pennsylvania, was the first to import them into the U.S. in 1900. The Sebastopol was never accepted as a commercial bird but are an attractive, different, novelty exhibition bird.

Sebastopol geese were recognized by the American Poultry Association in 1938 and placed in the breeding standards of England, Denmark, Holland, France and Germany.

Weights

Adult gander	14 pounds (6.35 kg.)
Young gander	12 pounds (5.45 kg.)
Adult goose	12 pounds (5.45 kg.)
Young goose	10 pounds (4.55 kg.)

Transylvania Naked Necks

Black Naked Neck Cock

This unique breed is typically known as Naked Necks due to the lack of feathers on its neck. Their history of origin has been lost. They may have had heritage in Japan because they were painted on Japanese paper fans. They are a very old breed and mentioned in a book of 1810, The *Treatise.* The name Turken applies to them as well as Churkeys, Turkey Necks, Gillikins and Nudes.

Naked Necks were widely distributed in the area of the Black Sea and known in Romania and Hungary. They came to England via Malta. Transylvania Naked Necks are ugly chickens but are exhibited and represented in the breeding Standards of Great Britain, Denmark, Germany, Holland, France and the United States. Naked Necks are considered fairly good layers of medium sized eggs. They are bred in White, Red, Buff , Black and were recognized in the *American Standard of Perfection* in 1965.

Weights

Cock	8 1/2 pounds (3.85 kg.)
Cockerel	7 1/2 pounds (3.40 kg.)
Hen	6 1/2 pounds (2.95 kg.)
Pullet	5 1/2 pounds (2.50 kg.)

Russia

Adler Silvers

The Adler Silvers are a variety developed from multiple crossings of Russian Whites and New Hampshires, Pervomayski and White Plymouth Rocks. They are used for egg and meat production. The average egg output is about 170 a year.

Weights

Cock	8 pounds (3.5 kg.)
Hen	6 pounds (2.7 kg)

Adler Silvers Trio

Orloffs

Sir Edward Brown reports seeing Orloffs in 1899 at the International Exhibition in Petersburg. He quotes Houdekow, '*To be a very ancient breed with likely ancestry of the Maley'.* Though they were known earlier in Russia, he mentions the possibility that their name came from Count Orloff Techesmensky who imported the breed from Persia (Iran). In Persia, they were known as the Giljanskya Kuriza chicken.

The Orloffs were first known in Russia as Chilianskaia chicken and are recognized as fairly good layers, very hardy and able to cope with the harsh Russian climate. They were imported into Germany in 1884 and England in 1899.

An article by Alexis Ossipoff states the Orloffs are a very old breed, descendants of hens brought from Persia. They were very fashionable until the introduction of other breeds like Cochins, Brahmas and Orpingtons. The Orloffs fell into decline and became obscure. Ossipoff asked the question, '*Why do we not protect the Orloffs, the breed of true Russian origin that*

Top
Cuckoo Orloff Hen

Bottom
Spangled Orloff Cock

White Orloff
Cock

are good layers and have fine meat quality and good beauty?' However, they continue to be obscure in Russia and they do not appear in the *Russian Standard.*

There are more Orloffs bred and exhibited in Germany than in any other country. They are also represented in the breeding Standard books of Great Britain, France, Denmark and Germany.

Weights

Cock	8 pounds (3.60 kg.)
Cockerel	7 pounds (3.20 kg.)
Hen	6 pounds (2.70 kg.)
Pullet	5 pounds (2.25 kg.)

Poltava Clay Colored

The Poltava is a local Ukrainian breed. The Ukraine Poultry Research Institute had done considerable work to improve egg production characteristics of the Poltava Clay Colored chicken, increasing it to 190 eggs a year.

Weights

Cock	6 pounds (2.7 kg.)
Hen	4 1/2 pounds (2.0 kg.)

Poltava Clay Colored
Trio

Ukraine Crested

The Ukraine Crested is a local breed for courtyard keeping. This ancient breed was reported by Sir Edward Brown in his book of 1906, *Domestic Races of Poultry.*

Weights

Cock	6 pounds (2.70 kg.)
Hen	4 1/2 pounds (20) kg.)

Ukraine Ushanki

The Ukraine Ushanki is identified by its beard. Nothing is known of their origin except that they were a popular fowl in the courtyards of the Ukraine and are a very old breed.

Weights

Cock	6 pounds (2.7 kg.)
Hen	4 1/2 pounds (2.0 kg.)

Top
Ukraine Crested
Cock

Bottom
Ukraine Ushanski
Hen

Top
Yurlov Golosysti
Trio

Yurlov Golosysti

The Yurlov Golosysti is known as a loud-voiced and long-crowing chicken. They are an attractive breed but have a low rate of egg production, about 120 a year.

Russian Fighting Geese

Arsamas

The Arsamas is named after the town of Arsamas. They were produced for goose fighting from a local race of geese by selective breeding. Some were used for utility purposes. The Arsamas is native to Gorky, and the population of this goose has not spread. They are kept by pet fanciers and lovers of the goose fighting sport. They have an all white plumage and range in size from 17 to 22 pounds.

Arsamas Fighting Geese

Kholmogory

The Kholmogory is a cross of Chinese geese with local domestic geese originally from central Chernozem regions. They roamed over Kursk, Byelgorod, Voronezh, Moscow, Vladimir and Arkhangelski districts. At present, they are raised in farmyards and state farms. The Kholmogory is the largest of the fighting geese.

They are distinctive because of the knob on their forehead but are not known for their meat production. Their plumage is Grey, White or Skewbold with weights ranging from 20 to 23 pounds.

Tula

The Tula is the best known of the fighting geese and are bred in the central districts of Tula, Kalonga, Tamboff, Kizan, Kourski, Vladimir, Njni-Novogorod, the upper borders of Oka, and the upper parts of the Volga River. The Tula is grey in color but some are clay colored. Their weights range from 13 1/2 to 17 pounds.

Top
Kholmogory Fighting Geese

Bottom
Tula Fighting Geese

Scotland

Scots Dumpy

The Scots Dumpy has been bred in Scotland for over one hundred years though poultry with similar characteristics has been described as early as the 17th century. Legend says the ancient Scots or Picts used them as time clocks. Sir Edward Brown quotes, '*They generally are supposed to have been brought to Scotland by the Phoenician traders long before the invasion of the Romans'.*

The Scots Dumpy distinctive feature is the shortness of their legs. For show purposes, they should not be more than 1 1/2 inches long. The short legs create an unique waddling gait like that of a duck. They are bred in Cuckoo, Black and White. Bantam Scots Dumpy are also bred.

Weights

Cock	7 pounds (3.20 kg.)
Hen	6 pounds (2.70 kg.)

Black Scots Dumpies Pair

Sicily

Sicilian Buttercups

The origin of the Sicilian Buttercups is a mystery. Reverend Ray Trudigan, England, considers them the chicken of Biblical times. They were imported into the United States in 1835. Mrs. Colbeck of Yorkshire was the first to import them into England in 1912.

Butercups were very plentiful in England in the 1960's. They are excellent layers of white eggs but are not a popular bird at shows. In the United States they are available commercially as baby chicks. Sicilian Buttercups are recognized in the *British Poultry Standard* and were accepted in the *American Standard of Perfection* in 1918.

Sicilian Buttercup Cock
Sicilian Buttercup Hen

Weights

Cock	6 1/2 pounds (2.95 kg.)
Cockerel	5 1/2 pounds (2.50 kg.)
Hen	5 pounds (2.35 kg.)
Pullet	4 pounds (1.80 kg.)

South Africa

Golden Australorp Cock
Golden Australorp Hen

Australorps

White Australorps originated from sports of Black Australorps imported from Australia. They were first exhibited by N.H. Addison, Middleburg Cape, and Dan Jacobs, Alberton. They were admitted to the *South African Poultry Standard* in 1939.

Golden Australorps were perfected by Professor A.M. Gericke of Pretoria University and admitted to the *Standard* in 1949.

Wheaton Laced Australorps were produced from sports bred from pure Australorps by Myrtle Phillips of Tarkastad and admitted to the *Standard* in 1954.

There is no development information on Blue Australorps.

Weights

Cock	8.8 pounds (4.0 kg.)
Cockerel	6.6 pounds (3.0 kg.)
Hen	7.0 pounds (3.2 kg.)
Pullet	6.6 pounds (3.0 kg.)

Natal Games

There is no information on the development of the Natal Games except that they originated in Natal. Allen Hope, a fancier, has an affidavit dated June 24, 1954, proving that he has bred true Natal Games since 1910 without the introduction of any other game fowl blood. This breed lays tinted eggs.

Weights

Cock	9 pounds (4.1 kg.)
Cockerel	7.4 pounds (3.4 kg.)
Hen	6 pounds (2.7 kg.)
Pullet	5 pounds (2.3 kg.)

Natal Game Cock

Natal Game Hen

Potchefstroom Koekoek Cock
Potchefstroom Koekoek Hen

Potchefstroom Koekoek

The Potchefstroom Koekoek was developed by C.L. Marais, Research Institute of Animal Husbandry, Potchefstroom. Breeding began in 1947 by crossing a Black Australorps male with White Leghorn females. The filial hens and cocks were mated to each other. Few of the progeny showed a striped plumage color. The striped feather-colored hens of the second and third filial progeny were crossed with the striped feather-colored Plymouth Rock cocks to obtain progeny homozygous for color.

All yellow skin progeny were selected and Potchefstroom Koekoek cocks were crossed with Black Australorps hens, to introduce as much Black Australorps blood as possible. The cross breed hens were mated with Potchefstroom Koekoek cocks. A few with yellow skin appeared. Each year, only yellow skin cocks were used until all breeding stock showed a striped color and yellow skin.

Weights

Cock	7.7 pounds (3.5 kg.)
Cockerel	6.6 pounds (3.0 kg.)
Hen	6.6 pounds (3.0 kg.)
Pullet	5.5 pounds (2.5 kg.)

Spain

Blue Andalusians

Blue Andalusian Cock

The Blue Andalusians are native to Andalusia, a province of Spain. Their precise history of origin is a mystery. They are related to the Minorcas, but some writer's think they are related to Spanish and Catalinas. Before England imported Blue Andalusians they were known as Blue Minorcas.

In England, they produced a similar fowl by crossing black chickens with white sports. Imported into England in 1846, they were first exhibited in 1853. The black edging on some of the feathers of Blue Andalusians distinguishes them from similar blue fowl like Lebryana Blues and Blue Minorcas. It's believed they resulted from the crossing of Black Minorcas with Black Spanish sports.

Typical of most blue fowl, only one third will be bluish-slate gray, one third will be black and one third will be white or smoky white. In the United States, Arthur Pfeil, probably the only Blue Andalusians specialist in the world, bred these birds true to color for 20 years. Even though others who bought his stock tried to emulate him, they never succeeded. His secret was having established lines that when crossed produced 100 percent blue. The Blue Andalusians are remarkable layers of large chalk white eggs.

In Spain, there are Barred Andalusians, Black Andalusians and Partridge Andalusians. This breed was recognized in the *American Standard of Excellence* in 1874.

Weights

Cock	7 pounds (3.20 kg.)
Cockerel	6 pounds (2.25 kg.)
Hen	5 1/2 pounds (2.50 kg.)
Pullet	4 1/2 pounds (2.00 kg.)

Buff Catalinas

Buff Catalinas

The Catalinas came from the district of Prat, near Barcelona, Spain. They were known as Moorish fowl and first exhibited at the World Fair in Madrid in 1902. In the *American Standard of Perfection,* they are referred to as the Catalina Del Prat Leonada. Information from Spain indicates the Catalina and the Prats are one and the same fowl.

Minorcas played a part in their development. They were popular in 1904 when 5,000 were known to be in the region of Castile. The *German Poultry Standard* refers to them as the Kastilainer, but the *British Poultry Standard* doesn't acknowledge them. They were recognized in the *American Standard of Perfection* in 1949.

Weights

Cock	8 pounds (3.60 kg.)
Cockerel	6 1/2 pounds (2.95 kg.)
Hen	6 pounds (2.70 kg.)
Pullet	5 pounds (2.25 kg.)

Leon

The exact origin of Leons is not known, but they are a very old breed reported by the monks in 1624. The feathers of Leon are highly prized for fly fishing lures. Two colors are known, the Indio de Leon and the Pardo de Leon. Feathers from the back are the most valuable but the hackle feathers are also used. The best Leons are raised in the Curueno River valley.

Breeders raised Leon in small flocks to supply a supplemental income and were

Pardo Leon Cock

secretive about how they bred them. They never sold hatching eggs, chicks or adults. The eggs were used only to provide food for their family. Feathers of the Leons are famous and marketed worldwide.

Weights

Cock	5.2 pounds (2.4 kg.)
Hen	3.9 pounds (1.8 kg.)

Minorcas

The Minorcas are the largest of the Mediterranean breeds. Details of their origin is unknown. At one time, they were called Black Spanish, Portugal Fowl and Red Faced Black Spanish. There is some evidence that they were from the Catalinas of Spain. This breed has been traced to 1780 on the Island of Minorca.

Black Minorca Cock

Moubray referred to them as a sub-species of the Spanish. They were first introduced into England in 1830 and Germany in 1870. Minorcas became very popular in England where breeders improved them. They are noted for their exceptional egg production as well as extra large egg size.

White Minorcas came to the United States in 1885. White and Black were recognized by the American Poultry Association in 1888. In 1900, George Northrup, Raceville, New York, developed the Rose

Black Minorca Hen

Comb Black Minorcas. They were the result of crossing Single Comb with Black Hamburgs. Rose Comb Blacks were accepted into the *American Standard of Perfection* in 1904. Single Comb Buff Minorcas were recognized in 1913 and Rose Comb White Minorcas were accepted in 1914.

Weights

Cock	9 pounds (4.10 kg.)
Cockerel	7 1/2 pounds (3.40 kg.)
Hen	7 1/2 pounds (3.40 kg.)
Pullet	6 1/2 pounds (2.95 kg.)

White Faced Black Spanish

Nothing specific is known about their origin or how the White Faced Black Spanish developed. They are considered one of the most ancient breeds and the oldest of the Mediterranean varieties. They were known in the Netherlands and England for centuries. Distribution was widespread including Egypt and Africa.

In the early days, the White Faced Black Spanish were a very common, popular bird in London. Three hundred were exhibited at the Birmingham Poultry Show in England in 1865. They are good layers of large, chalk eggs with an English breeder reporting 200 to 250 eggs a year. A pair of White Faced Black Spanish was exhibited in England in 1852.

In the United States, one breeder is known to own the White Spanish but they are not exhibited. They were acknowledged in the *American Standard of Excellence* in 1874.

White Faced Black Spanish Cock

Weights

Cock	8 pounds (3.60 kg.)
Cockerel	6 1/2 pounds (2.95 kg.)
Hen	6 1/2 pounds (2.95 kg.)
Pullet	5 1/2 pounds (2.50 kg.)

Sumatra

Black Sumatra Games

Black Sumatra Games were first imported into the United States from the Island of Sumatra in 1847, to England in 1849 and Germany in 1882. They were also known on the Islands of Java and Borneo where natives prized them for their cockfighting ability. They are called Pheasant Fowls, Pheasant Sumatra, Malaca Games and Sumatra Ebony Game chickens. Bennett, in his book of 1850, called them English Raven.

The Sumatra was widely exported and are liked by Japanese fanciers. They are flighty and extremely aggressive. One writer reported his birds are capable of flying across a 70 foot stream. Their beauty has special appeal to the poultry fancier because of their lustrous black-greenish sheen, distinctive Pea Comb and long tails. The sickle feathers can be 30 inches long. Another characteristic is their strutting, jaunty swagger. The White and Blue Sumatras were developed but never recognized in breeding standards. They are considered prolific layers and acknowledged in the *American Standard of Excellence* in 1894.

Black Sumatara Cock

Weights

Cock	5 pounds (2.25 kg.)
Cockerel	4 pounds (1.80 kg.)
Hen	4 pounds (1.80 kg.)
Pullet	3 1/2 pounds (1.70 kg.)

Sweden

Blue Swedish Duck

The most famous of the Swedish poultry is the Blue Swedish duck known to have existed as early as 1835. Surprisingly, it was the Germans who first showed them in Neuvor Pommern. The Dutch, Belgians and Germans import many because they are profitable due to their fast growth rate. Unfortunately, I have not been able to trace their origin.

Typical of most Blue fowl, they do not breed true to color. Only about one third will be Blue (Slate grey), with one third Black and one third a variation of White, Smoky White.

Weights

Drake	8 pounds (3.60 kg.)
Young drake	7 pounds (3.20 kg.)
Duck	7 pounds (3.20 kg.)
Young duck	6 pounds (2.70 kg.)

Blue Swedish Drake

Oland Geese

Oland geese are relatively rare in Sweden and it has been impossible to obtain any information about their origin. Oland is an island to the south of Sweden, however, there are no Oland Geese on the island. They are in the *Danish Breeding Standard.*

Weights

Gander	11 pounds (5.0 kg.)
Goose	9.9 pounds (4.5 kg.)

Oland Gander

Queen Silvias

I met Dr. Martin Silverud, a geneticist, in 1984 at the World Poultry Congress in Helsinki, Finland, when he would have a presentation before the Congress. He had developed a new breed of chicken, Queen Silvias, that laid eggs in a variety of colors.

Currently, it is unknown how many Queen Silvias are in Sweden. This is a chicken that can be auto sexed. The egg colors are similar to Araucanas. They are bred in Silver and Golden. I do not have weights on them but they are the same size as Leghorns.

Queen Silvia Pullet and varied colored eggs

Switzerland

Appenzeller Spitzhauben

Information from the Appenzeller Spitzhauben Society in England indicates they are an ancient breed brought to northern Europe by the Vikings. They were developed in the 1900's from crested stock in Switzerland. Brabants may have been crossed to bring about the unusual crests. The name originated from an area of Switzerland famous for its lace, Canton Appenzell, and Spitzhauben refers to the white spotted bonnet on their head.

They were exhibited before 1930 as Hornets. Pamela Jackson brought the first Appenzeller Spitzhauben to England in 1973 where they were so popular a specialty club was formed. They are bred in Silver Spangled, Gold Spangled and Black and are recognized in the breeding Standards of England, Denmark, Germany and Switzerland.

Weights

Cock 4 pounds (1.8 kg.)
Hen 3.3 pounds (1.5 kg.)

Top
Silver Appenzeller Spitzhauben
Cock

Bottom
Black Appenzerller Spitzhauben
Hen

Taiwan

Incubation Procedures

Taiwan enjoys a mild climate and almost any building can be used as an incubator. To start the procedure, eggs are preheated, either with a charcoal stove or electric heat. Initially, this also involves the use of heated rice bran. Operators work and live within the house. Checking the egg temperature is done by holding it to the eye lid with precise expertise.

The eggs are placed in a gunny sack, 90 eggs per bag and turned twice a day. They are candled two or three times during the early incubation period to eliminate infertile eggs. Cracked eggs, a result of turning, are also removed.

After the sixth day, eggs are transferred to egg beds. Eggs in this later stage of incubation are placed next to eggs in stages of development which creates a heat source. Hatches from these crude incubators compare favorably with expensive incubators. This method of incubation is a business in Taiwan and is done with a custom hatching arrangement with the owner, guaranteeing a 75% hatch rate.

Buff Tsai-Ya Drake

Buff Tsai-Ya Duck

Duck raising in this country dates back to the Ming Dynasty. One of the leading breeds is the Buff Tsai-Ya. They came from China in ancient times and vary in color. The White Tsai-Ya was developed for the Japanese market. They are bred for their exceptional egg laying ability, with an average of 250 eggs a year, with some laying 275 to 300. A desirable quality of Tsai-Ya eggs is that 90 % will hatch.

The Buff Tsai-Ya lay green, bluish and white eggs. The green and blue command premium prices because they stand out from other eggs. They are sent to select markets and are considered a delicacy.

Buff Tsai-Ya Duck

The Tsai-Ya is closely related to Mallards and the males are crossed with Muscovy ducks to produce a species called Mule ducks. Duck breeders in Taiwan have been skillful in their production methods efficiently utilizing artificial insemination. Because the people of this country consume many duck eggs, high production levels of eggs and meat is essential.

Weight
Duck 3 pounds (1.40 kg.)

Chickens in Taiwan have less importance than ducks. Like most nations, the dominant egg supply is furnished by large commercial egg facilities usually from a hybrid cross. Breeds are known as Country Chickens or Native Chickens.

Native country chickens of Taiwan

Turkey

Denizili

In Denizili this breed is prized for its long-crowing ability. The exact origin of the Denizili isn't known but they are of great importance to the people of Turkey and are propagated by the Turkish National Agricultural Institute for distribution to farmers. In the city of Denizili there is a large rooster statue in honor of them.

The crow of an average rooster is only a few seconds, but the crow of Denizili can range from 20 to 20 seconds, with a record crow of 45 seconds.

Weights

Cock	7.7 pounds (3.50 kg.)
Hen	5.5 pounds (2.50 kg.)

The Denizili
The long crowing chicken of Turkey

Statue in honor of the Denizili in the city of Denizili

Sultans

The Sultans are considered rare in Turkey and were known as the *Serai-Taook*, meaning *Fowls of the Sultan.* They may have been the result of a White Polish cross with White Booted Bantam. Elizabeth Watts, England, imported the first Sultans from Constantinople in 1854. The British improved the original stock by crossing them with White Polish.

Sultans were imported into the United States in 1867 and were accepted in the *American Standard of Excellence* in 1874. They are in the Standards for Germany and England.

Weights

Cock	6 pounds (2.70 kg.)
Hen	4 1/2 pounds (2.00 kg.)

White Sultan Hen

White Sultan Cock

United States

Black East India

I am including this duck in the United States section because their country of origin is a mystery. Paul Ives writes in *Domestic Ducks and Geese*, that they may have been derived from Black ducks in Europe. Sir Edward Brown wrote they may have been involved in the first breeding that developed the Cayugas. In England, they were included in the *Book of Standards* n 1865.

In ancient poultry history the Black East India was known as Black Brazilian and Buenos Aires duck. Darwin referred to them as Labrador duck.

They are a very popular Bantam duck and recognized in the *British Breeding Standard*. The Black East India was accepted in the *American Standard of Excellence* in 1874.

Black East India Drake

Weights

Drake	30 ounces
Young drake	26 ounces
Duck	24 ounces
Young duck	22 ounces.

Buckeyes

Buckeye Cock
by
Mary Ellen Johnson

The Buckeyes were originated by Nettie Metcalf, Ohio, by breeding Cornish (Indian Games), Brahma and Black Breasted Game Fowls. The Pea Comb came from the influence of Brahma and Indian Games. Due to their similarity to the Pea Comb Rhode Island Red, (developed at the same time) the American Poultry Association would not recognize them immediately. Rhode Island Red breeders decided to drop their Pea Comb Reds.

The popularity of the Buckeye grew and a specialty club was established. However, over the years it was Rhode Island Reds that gained in popularity. Buckeyes were accepted in the *American Standard of Perfection* in 1905.

Weights

Cock	8 1/2 pounds (3.85 kg.)
Cockerel	7 1/2 pounds (3.40 kg.)
Hen	8 pounds (3.60 kg.)
Pullet	5 1/2 pounds (2.50 kg.)

Cayuga

Cayuga Drake

J.S. Clarke, Cayuga County, New York, is credited with the development of this duck but Lake Cayuga inspired the name. They were called to public attention by a newspaper article in 1851 but nothing was said about their development as a breed. Sir Edward Brown wrote that large species of Black East India ducks may have had a place in the heritage of the Cayuga.

The Cayuga is a fairly popular exhibition duck in the United States and was introduced into England in 1851. They were recognized in the *American Standard of Excellence* in 1874.

Weights

Drake	8 pounds (3.60 kg.)
Duck	7 pounds (3.20 kg.)

Delawares

Delawares Pair
by
Mary Ellen Johnson

Credit for the development of Delawares goes to George Ellis, Ocean View, Delaware. He was assisted by Dr. Edmund Hoffman, Canada. In 1940, lighter colored sports from a Barred Rock and New Hampshires broiler cross were selected to breed. It was not intended to develop a new breed, but rather to establish a male line with a Colombian pattern. This was done to avoid the dark pin feathers found in the broiler cross.

When this was accomplished Delawares were marketed as Indian Rivers in the commercial broiler industry and were popular for awhile. Though they are still available it is apparent the broiler industry prefers the Cornish Rock cross.

Delawares were never popular with exhibitors. The Delaware Bantam was originated by Cecil Moore, Texas. Delawares were recognized in the *American Standard of Perfection* in 1952.

Weights	
Cock	8 1/2 pounds (3.85 kg.)
Cockerel	7 1/2 pounds (3.40 kg.)
Hen	6 1/2 pounds (2.95 kg.)
Pullet	5 1/2 pounds (2.50 kg.)

Trio of Dominques

Dominques

Dominques were reported as far back as 1750 and are recognized as the oldest breed in the United States. Bement wrote about them in 1856, '*Generally looked upon as a mere farmyard fowl that is the accidental result of promiscuous crossing*'.

The Dominques character, type and refinement came when exhibiting began. One reference reports their similarity to the Cuckoo Dorkings of England. Another early writer, McGrew, states '*We know the Dutch were among the earliest settlers in the New York area. They logically would bring their chickens from Holland. Their pencilled Hamburgs, that are Rose Comb, ties in with their intermingling with other fowl, so perhaps some basis for our Dominques.'*

Sir Edward Brown wrote that they were first brought over by the early Puritans. They were a popular breed in the first poultry show, November 1849, held in Boston. This fowl was recognized in the *American Standard of Excellence* in 1874.

Weights

Cock	7 pounds (3.20 kg.)
Cockerel	6 pounds (2.70 kg.)
Hen	5 pounds (2.25 kg.)
Pullet	4 pounds (1.80 kg.)

Jersey Giants

Jersey White Giant Cock

This exceptionally large breed was first called Black Giants and originated in 1887 by John and Thomas Black. The crosses to develop them were Black Langshans, Dark Brahmas, Cornish and Black Javas. U.L. Meloney became the Black Giants patron and bred them at Marcy Farms, New Jersey. He exhibited extensively in an effort to popularize them in the U.S. As a result of his efforts, the Marcy Giants became accepted by that name. Then, Elwin Sears, Ohio, bred them in 1923 and they became known as the Sears Jersey Giants. For years the Jersey Giants were the leading breed for capons. The Cornish Rock cross now dominates that market.

Jersey Black Giant Hen

Mrs. R.E. Miller, Kansas, developed the Blue and Splashed Jersey Giants and is the foremost breeder, exporting her strain all over the world. Despite their large size, they are fairly good layers of large brown eggs. Jersey Black Giants were accepted in the *American Standard of Perfection* in 1922 and Whites in 1947.

Weights

Cock	13 pounds (5.90 kg.)
Cockerel	11 pounds (5.00 kg.)
Hen	10 pounds (4.55 kg.)
Pullet	8 pounds (3.60 kg.)

Lamonas

Lamonas were developed by Harry M. Lamon, at the U.S. Government Experimental Staten, Beltsville, Maryland. The breeding started in 1912 and was done to perfect an improved dual purpose egg and meat fowl. Three breeds were used in producing Lamonas, Silver Grey Dorkings, White Leghorns and White Plymouth Rocks. It took 22 years of careful selection before type

Trio Lamonas

and characteristics established them as a new breed.

Lamonas are excellent layers and if trapnested will produce over 200 eggs a year. With a good record for large eggs and a nice, yellow skin, Lamonas had excellent prospects for the future. The Osborne Poultry Farm of Holland, Michigan, and S.E. Raymond, Ohio, promoted them extensively. They were exhibited at the World Poultry Congress in 1933 and accepted into the *American Standard of Perfection* in 1934. They are now on the SPPA endangered list.

Weights

Cock	8 pounds (3.60 kg.)
Cockerel	7 pounds (3.20 kg.)
Hen	6 1/2 pounds (2.90 kg.)
Pullet	5 1/2 pounds (2.50 kg.)

Trio New Hampshires

New Hampshires

This breed was developed for the broiler industry. Breeding work started in 1915 at the New Hampshire Agricultural Experiment Station. Rhode Island Reds were the only breed used in their development. The goal was to raise a fast-feathering bird with good growth rate, early sexual maturity and good-sized brown eggs. Timing of this program ties in with early blood testing for Salmonella Pullorum that began in 1916. This test involved drawing a blood sample from each bird to be tested to detect reactors to Pullorum. The New Hampshires were the first and perhaps only breed developed from a group that were proven free from Salmonella Pullorum.

New Hampshires are much lighter in color than Rhode Island Reds, but color was secondary. Priority was given to fast maturity and good livability. They are exhibited and color is standardized around the world. Hubbard Farms and Christie Farms of New Hampshire were breeders who made them popular in the U.S. The New Hampshires were acknowledged as a new breed in the *American Standard of Perfection* in 1935.

Weights

Cock	8 1/2 pounds (3.85 kg.)
Cockerel	7 1/2 pounds (3.40 kg.)
Hen	6 1/2 pounds (2.95 kg.)
Pullet	5 1/2 pounds (2.50 kg.)

Pilgrim Geese

Pilgrim geese are noted for sex-linked characteristics making it possible to visually determine the sex of a day old gosling. The ganders are white and the goose is grey. Oscar Grow, author of the book, Modern Waterfowl Management and Breeding Guide, was given credit for their development in 1930. He credits the name of Pilgrim to his own pilgrimage from Iowa to Missouri. England has similar sex-linked geese called West of England. In the *British Poultry Standard*, Pilgrim geese are listed but it's implied their development was in England. In Normandy, France, there is a species of geese that resembles the West of England.

An English poultry book gives credit to Pilgrims in 1620 for the introduction of these geese. There is a reference to auto sexed geese in Reverend Jeyn's book of

Pilgrim Geese Pair
Ganders are white, the goose, gray

1800, *Vertebrate Animals*. It may be there is no connection with the Oscar Grow development of the Pilgrim geese because they were apparently known in earlier days.

There is another contradiction in Paul Ives's book of 1947, *Domestic Ducks and Geese*. He states the Pilgrim went from the Netherlands to Massachusetts in 1620 and quotes Boswell's book of 1841 that said the ganders were white and geese grey. He also mentions M. Parmentier, a French writer, referred to geese with white ganders and grey females. This raises the question that perhaps these were the geese now called Pilgrim.

The Pilgrim is an excellent layer and there is a waterfowl hatchery in the U.S. that produces just Pilgrim goslings. The hatchery is the Pilgrim Goose Farm of Williamsfield, Ohio. They have a breeder flock of over 200. The Pilgrim has not proven popular as exhibition birds. They were recognized by the *American Standard of Perfection* in 1939.

Weights

Gander	14 pounds (6.35 kg.)
Young gander	12 pounds (5.45 kg.)
Goose	13 pounds (5.90 kg.)
Young goose	10 pounds (4.55 kg.)

Barred Plymouth Rocks Pair

Plymouth Rocks

Barred

Several breeders and much experimentation was involved with the development of the first Plymouth Rocks. It has been established that in 1847 there was an intermingling of several breeds including White Cochins, Dark Brahmas, White Brahmas, Black Javas, Langshans, Dorkings, Black Minorcas, Black Spanish and Dominques. However, three Plymouth Rocks were exhibited at the first poultry show in Boston in 1849 by Dr. John Bennett despite the fact they were far from being perfected.

In a special edition of *Plymouth Rock Standard* published by the American Poultry Association in 1915, there is information that is contradictory to most historical knowledge of the Plymouth Rocks. This book states the first Rocks became extinct in 1811. From this, I conclude there were earlier efforts to establish this new breed. Dr. Bennett, Massachusetts, is credited with being the originator of the Plymouth Rocks, developing them from a cross of a Cochin China hen with a Fawn Colored Dorkings. Upham is given credit for producing the Barred Plymouth Rock by crossing a Dunghill cock with a Black Cochins hen.

Harrison Weir wrote, '*The Plymouth Rock Fowl is then in reality one half Cochin China, one fourth fawn colored Dorking, one eighth Malay and one eighth Wild Indian.'* In one instance, they were reported to have been improved with the Dominques.

There are references to different strains of Plymouth Rocks including, Gilman, Essex County, Spaulding, Ramsden, Drake and the Upham. Breeders mentioned in their development are, Dr. Bennett, Ramsden, Robinson, Shedd, Spaulding, Gilman, Essex, Lord, Drake, May, Gray and Upham.

In March, 1869, Plymouth Rocks were exhibited by D.A. Upham, Worcester, Massachusetts. Parks Poultry

Farm, Altoona, Pennsylvania, who specialized in this breed commercially and made them well known in the United States.

Though Plymouth Rocks are layers of nice brown eggs they are equally significant for their excellent meat qualities. They are recognized in the *American Standard of Excellence* in 1874.

Weights

Cock	9 pounds (4.10 kg.)
Cockerel	7 1/2 pounds (3.40 kg.)
Hen	7 pounds (3.20 kg.)
Pullets	6 pounds (2.70 kg.)

Blue

I found no recorded history pertaining to the development of Blue Plymouth Rocks. It is interesting that there would be Black Plymouth Rocks recognized in the *British Poultry Standard.* Some Buff Barred Plymouth Rocks were developed as well as a Rose Comb variety. However, neither of these survived the test of time. Blue Plymouth Rocks were admitted to the *American Standard of Perfection* in 1920.

Weights

Cock	9 pounds (4.55 kg.)
Cockerel	7 pounds (3.40 kg.)
Hen	7 pounds (3.20 kg.)
Pullet	6 pounds (2.70 kg.)

Buff

There is no relationship shown between Buff Plymouth Rocks and Barred Plymouth Rocks.

At the World's Fair in 1893, they were exhibited as the Wilson Strain. This was a cross between Buff Cochins and Light Brahmas. In 1890, Fall River Strain was exhibited by R.G. Buffington, Providence, Rhode Island, and involved Rhode Island Reds. There is also reference to English Buff Rocks that included blood of Buff Orpingtons.

Sir Edward Brown reported another combination of crossings Buff Leghorns, Buff Cochins, and Light

Brahmas. I think Buff Plymouth Rocks are a blend of all breeding efforts. They were accepted into the *American Standard of Excellence* in 1894.

Colombian

The Colombian Plymouth Rocks were known to have developed in Ohio. McGrew states in *The Book of Poultry* there was a mating of a Light Brahma colored male with the shape of Plymouth Rocks, to Barred Rocks and White Rocks to attain the Colombian pattern. Another strain was a cross of Light Brahmas with White Plymouth Rocks, then re-crossing the offspring with males of other strains.

The Colombian Wyandottes played a part in their development. The Colombian Plymouth Rocks were recognized as a new breed and admitted to the *American Standard of Perfection* in 1910.

Partridge Rocks

There have been a number of strains of Partridge Rocks developed. The first originated at the Elmwood Farms, New Jersey. They were the result of offspring from Partridge Wyandottes. Another strain established in Indiana was a cross of Dark Cornish or Indian Games male with Partridge Cochins female and the offspring of this, mated with Golden Laced Wyandottes. Top breeders were selected to be re-mated with the fowls to attain the character of Plymouth Rocks with Partridge color.

The Partridge Plymouth Rocks were the result of the blending of these strains. Brown Leghorns were reported as part of their breeding. They were recognized as a new breed in the 1909 *American Standard of Perfection.*

Silver Pencilled

Developed in New York in 1894, the offspring of the Silver Pencilled Wyandottes with Single Combs is known to be the starting point for the Silver Pencilled Plymouth Rocks. Other strains were developed by mating Silver Grey Dorkings with Dark Brahmas. The best of the offspring was bred until the present type was produced. They were admitted to the *American Standard of Perfection* in 1907.

White

William Shedd, Waltham, Massachusetts, played a part in the development of the first Plymouth Rocks. He started breeding them in 1847, from sports of the Plymouth Rock. There is another report that they are the result of crossing of a Light Brahmas female with a White Dorkings cock. In 1874, Oscar Frost, Monmouth, Maine, also helped to develop the first White Rocks after successfully hatching a few chicks from Barred parents that were all white. They were recognized by the American Poultry Association in 1888.

White Rock Cock

Rhode Island Reds

In Professor John Skinner's book, *American Poultry History*, he states that the start of this breed was in 1830 in Narraganset Bay, Rhode Island. In a 1954 article by David Patten, he reports it all began in 1854 with a crossing of a Malay or Chittagong crossed with scrub farm hens. He mentions the cross to be a Malay with Dark Brahmas and credits his grandfather, Isaac Wilbour and William Tripp with their development. The article states the Tripp Fowl were better meat birds and better layers. The Tripp Fowl was crossed with Plymouth Rocks, selecting the best Buff hens which were then mated with the best Red roosters. It was at this point Dr. A. Birnham and Samuel Cushman, U.S. Experiment Station , Kingston, Rhode Island, took an interest in them. Isaac Wilbour was the first to advertise Rhode Island Reds and is credited with naming them.

Rhode Island Red Cock

Sir Edward Brown stated they were first exhibited by Jenney, 1879, at the South Massachusetts Poultry Show. In 1891 they were exhibited at the Philadelphia Poultry Show under the name of Golden Buffs. It is obvious that color had not yet been established. Brown mentions the names of several breeders but doesn't pinpoint whom developed this new breed.

In 1925, the only monument ever to be erected in honor of a chicken

Rhode Island Red Hen

(Rhode Island Red) was commissioned in Adamsville, Rhode Island. The Rhode Island Red is the only chicken to be recognized by television with a 100 year Anniversary program. The State of Rhode Island published a book to recognize them and credits their development being Little Compton, Rhode Island.

Practical qualities were foremost in producing Rhode Island Reds but not for exhibition. Trapnesting became popular and these chickens had the best laying record, up to 250 eggs a year, with some topping 300. They laid brown eggs which were popular and purchased by commercial outlets. Rhode Island Reds were often called the Leghorns of the Heavy Breeds.

Exhibitor breeders developed a deep Mahogany Red color which was passed up by commercial breeders who felt the exhibition birds weren't as good layers as commercial ones. Paramenter Reds were well known commercially but were not exhibited. The Single Comb Rhode Island Reds were recognized by the American Poultry Association in 1904 and the Rose Comb in 1905.

Weights

Cock	8 1/2 pounds (3.85 kg.)
Cockerel	8 pounds (3.60 kg.)
Hen	6 1/2 pounds (2.95 kg.)
Pullet	5 1/2 pounds (2.50 kg.)

Six foot high monument to honor Rhode Island Red in Adamsville, Rhode Island

Rhode Island Whites

Alonzo Jacoy, Peace Dale, Rhode Island, is credited with the development of Rhode Island Whites. He wanted to raise a breed that would suit a commercial market and crossed Cochins with White Wyandottes, breeding the offspring back to Cochins. In 1900, he completed Rhode Island Whites by adding Rose Comb Leghorns to the bloodline. There is no mention of Rhode Island Red blood being utilized.

Rhode Island Whites were offered to the public in 1903 but not accepted into the *American Standard of Perfection* until 1922. They never attained popularity.

Weights

Cock	8 1/2 pounds (3.85 kg.)
Cockerel	8 pounds (3.60 kg.)
Hen	6 1/2 pounds (2.95 kg.)
Pullet	5 1/2 pounds (2.5 kg.)

Rhode Island White

Wyandottes

Black

Credit goes first to F.M. Clements, Ohio, for the first Black Wyandottes. However, F.A. Marshall also contributed to their development which was the result of sports from Silver Laced Wyandottes. Black Wyandottes were recognized by the American Poultry Association in 1888.

Blue

There is no recorded history about their development. They were accepted in the *American Standard of Perfection* in 1977.

Weights

Cock	8 pounds (3.60 kg.)
Cockerel	6 1/2 pounds (2.95 kg.)
Hen	7 pounds (3.20 kg.)
Pullet	5 1/2 pounds (2.50 kg.)

Colombian

Reverend B.Briggs is credited with the development of Colombian Wyandottes. Breeds involved to produce them were White Wyandottes, Plymouth Rocks and Light Brahmas. They were first exhibited in 1893 at the Columbia Exhibition or World's Fair. Colombian Wyandottes were accepted into the *American Standard of Perfection* in 1905.

Golden Laced

The first Golden Laced Wyandottes were developed by Joseph McKeen, Wisconsin. Several breeds were involved, Silver Laced Wyandottes, Pea Comb Partridge Cochins, Brown Leghorns, Buff Cochins, and Golden Seabrights. Sir Edward Brown reported Winnebagoes were involved with Silvers, then re-crossed. They were acknowledged as a new breed in 1888 in the *American Standard of Excellence.*

Partridge

Two breeders played a part in the development of the Partridge Wyandottes, George Brackenbury, New York, and Ezra Cornell. Both men used the same breeds, Golden Laced Wyandottes, Golden Pencilled Hamburgs, Partridge Cochins, Cornish, Rose Comb Brown Leghorns and the extinct breed, Winnebagoes. Partridge Wyandottes were accepted as a new breed by the American Poultry Association in the *American Standard of Excellence* in 1901.

Silver Laced

The first of the Wyandottes family was Silver Laced. In 1865, John P. Ray, Hemlock, New York, is credited with their development. In 1870, Whitaker, of Michigan, bred fowls with silver hackles. These two strains were the start of the new breed, Silver Laced Wyandottes. For awhile, these chickens were called American Seabrights. There was a dispute about the name, possible presentations were Hambrights and Hambletonians. A compromise was reached when F.A. Houdlette, Massachusetts, called them Silver Laced Wyandottes, the name of his father's ship, the *Wyandotte.*

Some people believed the name was derived from the Wyandotte Indians. From 1877 to 1883, more crosses were made before establishing type and color uniformity. They were recognized in the *American Standard of Excellence* in 1883.

Silver Laced Wyandotte Hen

Silver Pencilled

There is no history regarding the Silver Pencilled Wyandottes nor the breeders involved in their development. They were accepted into the *American Standard of Perfection* in 1905.

White

The White Wyandottes appeared as early as 1872 and were developed from sports of the original Silver Laced Wyandottes. They were recognized as a new breed by the American Poultry Association in 1888.

White Wyandotte Cock

Commercial Hybrid Chickens

Scientifically, commercial hybrid chickens are truly a great accomplishment. The first commercial hybrid were perfected by the Pioneer Seed Corn Company of Iowa who sold their first hybrid chicks in 1942. They were marketed under the Hyline label by tremendous sales organization consisting of farmer salesmen. The Dekalb Seed Corn Company of Illinois hatched their first hybrid chicks in 1945. They, like many others entering the new era of the poultry industry, would offer various lines, identified by numbers such as, Dekalb 289 or Dekalb 307. These advances involved scientific selection, advanced record keeping and trap nesting. As computers came into increased use, they would play a strong role in assisting scientfic selection to identify the most efficient layers. Even commercial meat chickens for the broiler world would be developed using these highly scientific methods. Without the advancements of these major agricultural organizations, low priced eggs and poultry meat would not have been possible.

For many years, the requirement for eggs and poultry meat was supplied by mixed farmers and specialized poultry farmers. By the early 1920's we would see the start of the commercial hatchery industry supplying independent farmers with the requirement for baby chicks. It was during this period of time that my parents were to become pioneers in this new, fast growing hatchery idustry. My father and mother had been grain and cattle farmers and in 1924 they left their farm to build a modern chick hatchery in Fort Dodge, Iowa. In time, their hatchery was to become one of the larger hatcheries in the United States. At peak production, they hatched up to 2.5 million chicks and 250,000 turkey poults per year. As the demand for both eggs and poultry meat grew in the United States, so did the demand for poultry chicks. During the period through 1940 to 1960 there were a total of over 10,000 hatcheries operating in the United States.

Concurrent with the development of hybrid chicks, came the development of large scale commercial egg factories. Within a short period of time, these agricultural corporations would replace the eggs supplied by both farmers and small specialized poultry farms. Their advantage was the ability to produce a better and more uniform egg at a much lower cost per dozen. In the beginning, most of these egg factories were operated by single families with as many as 15,000 to 25,000 layers. Eventually, the big agricultural corporations would have farms with layers numbering in the millions. In time, these large egg factories privailed. In the United States today, the greatest share of the egg market is supplied by these giant agricultural corporations. This trend has also been true in most of the industrialized nations across the world.

Yugoslavia

Commercial Chickens

I am representing Yugoslavia to show commercial chickens. This country does not have any exhibition breeds for hobbyists. The Brown Prelux-R is auto sexed, considered a good layer at about 268 eggs per year, and weighs an average of 2.23 kg. The Barred plumage layer is indicated as Prelux-G and known to be a good layer, producing about 274 eggs a year and weighing 2.68 kg. The White chickens referred to as Prelux-BRO is a cross breed developed for meat. They are not good layers with only 156 eggs per year. The speed of feathering makes it possible to distinguish their sex early because pullets feather more quickly than cockerels.

Top to Bottom
Prelux-R auto-sexed
Prelux-G
Prelux-Bro

Zimbabwe

Shona Fowl

Very little is known about the Shona. They are thought to have come south with the Bantu tribes when they migrated from the west and north African regions. It's obvious these birds have a relationship with Translyvania Naked Necks, which were introduced by Arab slave traders. The difference is in the structure of the eye area. This identifies the Shona as a separate breed from the Naked Necks.

Due to their small size, Shona Fowl popularity waned with the introduction of larger, more popular breeds.

Shona Fowl Cock

All of the birds in the following section are not recognized in any recognized standards. In most cases the picture with no description is all we had. We present them for their interest value and have attempted to identify country of origin. Enjoy!

Canary Islands Cock

APPENDIX I
Associations and Specialty Clubs

Australia

Australia Pit Fowl Club of Australia: E.P. Vaughan
Elizabeth Street
Raleigh
New South Wales 2454

Orpington Club of Australia: R. Laverder
Wirrimbirra
Laheys Creek 2852

Old English Game Fanciers Association of Australia: W. Flint
58 Taralga Road
Coulburn 2580

Waterfowl Club of Australia (South Australia Branch): Don Bubner
P.O. Box 1
Owen

Macquarie Poultry Club: Judy Toynton
130 Pierce Street
Wellington
New South Wales 2820

Rose Comb Bantam Club of Australia: L. Cosfer
19 Brolgar Road
Pavkes
New South Wales 2870

Bantam Club of New South Wales: Rod Harris
291 Galston Road
Galston
New South Wales 2159

Exhibition Poultry Association of New South Wales: Rod Harris
291 Galston Road
Galston, New South Wales 2159

Ancona Club of Australia: M.A. O'Conner
Box 219
Murwillanbach 2484

Pekin Bantam Club of New South Wales: D.J. Plant
46 Newcastle Road
East Maitland
New South Wales 2323

Wyandotte Club of Victoria: R.J. Sharpe
43 Pirces Highway
Pakerham
Victoria 3810

Borneo Cock
Hen top left

Belgium

Oriental Game Fowl Club: Johny Van Impe
Molendijk 88
9550 Herzele

Het Neerhof Gent: Antoine Taghon
Mariakerkesteenweg 15
9031 Gent Drongen

Het Neerhof Voor Het Land Van Aalst:
Maurice Van DePette
Elenestraat 146
9620 Zottegem

Limburg Erf: R. Claes
Bverzakbroekweg 95
4300 Waremme

La Royale Hesbaye Avicole De Waremve: C. Geradon
Rue Delchambre 9
4317 Borlez

Brazil

Associacao Brasileira Dos Criadores: Perrone Pasquale
Rus General Lecor 872
04213 Sao Paulo

Canada

Canadian Araucana Society
Box 345
Sackville, Nova Scotia B4C 2T2

Canadian Bantam Club: Don Miedema
Route 7
Simcoe, Ontario N3Y 4K6

Manitoba Poultry Breeders Association: Mrs. John Pilchar
39 9th Street North
Brandon, Manitoba R7A 2R5

North Central Alberta Poultry Association: Bunny Felgate
217 Scona Glen Estates
23030 Twp. Road 522
Sherwood Park, Alberta T8B IHI

Ontario Poultry Assciation: Mrs. Cavell Ridler
Route 2
Guelph, Ontario NIH 6H8

Prince Edward Island Poultry Association: James Butler
Route 1
Montaque, Prince Edward Island COA IRO

Quebec Poultry Association: Emile Boucher
10455 Curotte Street
Montreal, Quebec Y2C 2YB

Saskatchewan Poultry Association: Mariane Ronsman
179 Coteau Avenue
Weburn, Saskatchewan S4H OG2

Vancouver Poultry Association: Ernie Silveri
2736 Lefeuvre Road
Aldergrove, British Columbia VOX IAO

Denmark

Danmarks Fjerkraeavlerforening for Raceavl
5260 Odense S.
Stenlosevj 292

Danske Landhens Club: Tom Nyby Thomsen
Storstraede 1
2620 Albertslund

Hjorring Poultry Club: Kaj Pedersen
Bredningen 108
Manna
9700 Bronderslev

Czechoslovakia
Hen

England

Rare Breed Society: R. Billson
Alexandra Cottage
8 St. Thomas Road
Great Glens, Leics

Sebright Club: J.K. Sharpe
Lynwood
37 Bowgeave
Garstang, Preston

Silkie Club: Mrs. D. Bryer
Lower Vicarwood Farm
Mackworth, Derbyshire

Sussex Poultry Club: Mrs. M. Ralsey
Exbury
Ashford Hill Road, Headley
Near Newbury, Berks

Welsummer Club: Mrs. Bullen
Armeria Sandford Land
Old Leake
Boston, Lincs PE229RB

World Poultry Science Association
Box 63, Westbury House
Burry Street
Guildford, Surrey GU2 5BH

Appenzeller Club: Ms. D. Smile
Oak Drive Lodge, Bromsberrow
Near Ledbury, Herefordshire HR8 IRY

British Araucana Club: Mrs. R. Williams
Bloomfield Cottage, Oakhill
Near Bath, Somersett BA3 53B

British Barnvelder Club: B. Clarke
113 Collingwood Road
Sutton, Surrey

British Faverolle Society: Mrs. S. Bruton
Park House, Codsall Wood
Wolverhapton, Staffs

British Waterfowl Association: Mrs. C.J. Winskill
6 Caldicott Close
Winsford, Cheshire CW7 I14W

Cochin Club: Michael Hatcher
Lime Tree Cottages
Brightwalton
Newbury, Berks RG16 OBZ

Croad Langshan Club: Mrs. C. Hadley
Heather Cottage, Deadmans Lane
Goring Heath
Reading, Berks RG8 7RX

Dorking Club: D. Chantry
Megrell Langley
Kirk Langley, Derby

Frizzle Society: Mrs. E. Amos
Penrhiw Cottage, Beuladh
Llanwtyd Wells
Powys, Wales

Leghorn Club: B. Turner
1 Oak Cottages, Barnham Lane
Walderton, W. Sussex

Marans Club: Mrs. A. Rolls
East Dunley, Dovey Tracy
Newton Abbot
Devon, England TQ13 9PW

Old English Game Club: Dr. H. Bell
Long Marston Manor
York, England YO5 8 LS

Orpington Club: Mrs. J. Bridson
4 Arhog Road
Didsbury
Manchester M20 OHD

France

Societe Centrale d_Aviculture de France
34 Rue de Lille
75007 Paris

Bantam Club Francais: J.C. Martin
G Rue Genevieve Plailly
91330 Yeeres

Groupement National des Eleveurs deBresse-Gauloise: E. Tamburini
19 Rue Cuzin
69120 Valulx en Velin
or
G. Jourdan
Chantereine, Lacelle Sur Loire
58440 Myennes

Indonesia Hen

Club de Volailles: J. Etienne
6 Rue Du Kem, Manom
57100 Thionville

Houdan - Faverolles Club: D. Duchange
9 Rue de Vernevil
75007 Paris

Marans Club: P. Bougrain-Dubourg
56 Fief De Montamer, La Nove
17740 Sainte Marie De Re

La Fleche Club: B. Briaud
Clos Du Sage Hardi
Allas
33124 Auror

Gatinais Club: Mr. Barbier
8 Rue Sermonnoise
77380 Combs La Ville

Brahma-Cochin Club: J.L. Vanlanghenhoven
7 Rue Ernest Roche
75017 Paris

North Korea Pair

Club Pour La Sauvegarde des Races Avicoles Due Poitou: P. Pichard
Pavillion Rhin Et Danube
Brain Sur L'Authion
49800 Trelaze

Cub Des Palmipedes Domestiques: J. Fernandez
10 Rue Longsart
Esnew
59127 Walincourt-Selvigny

Aviornis-France: J.P. Rolland
16 Rue Therese
33560 Carbon-Blanc

Gasconne Club: C. Raoust
37 Rue Joseph Marginal
Saint Martin Due Touch
31300 Toulouse

Groupement Zoologique Et Avicole: G. Dussart
62 Rue Des Sarts
59870 Rieulay

Bourbonnais Club: Mr. Benoit
Les Renauds
Paray Le Fresyl
03230 Chevagnes

Club Du Barbu De Watermael: J. Besselievere
7B Rue Jean Moulin
La Sanssaie
27370 Amfreville La Champagne

Club de la Volaille Meusienne: Jean Claude Periquet
Gincrey
55400 Etain

Germany

SV d.Z. des Barnevelder - und Zwerg-Barnevelder-Huhnes: (Barnevelds) Herman Hub
Mindenerstr 172 a
4900 Herford

SV d.Z. des New Hapshire-Huhnes: (New Hampshires) Friedhelm Schweickert
Amselstr. 18
6927 Bad Rappenau

SV d.Z. des Rhodelanderhuhnes: (Rhode Island Reds) Heinz Moller
Immenweg 30
2850 Bremerhaven

SV d.Z. Seltener Huhnerrassen: (Rare races of all ages) K.H. Forschler
Tannenstr. 12
7534 Birkenfeld 1

Club Deutscher ur-und Kampfhuhnzuchter: (All large fowl) Franz Messner
Hoppenriedweg 9
8940 Memmingen

SV d.Z. des Zwerg-Italiener-Huhnes: (Anconas, Leghorn Bantams) Heichrich Buschsieweke
Putzhagen 73
4830 Guterloh 1

South Korea
Cock

SV d.Z. der Zwerg-Langschan: (Langshan Bantams) Fred Sandmann
Stettinerstr 25
2808 Syke-Barrien

SV d.Z. der Zwerg-Niederrheiner: (Niedrrheiner) Herbert Krause
Mittelkamp 2
2740 Bremervorde

SV d.Z. der Zwerg-Orpington: (Orpington Bantams) Eberhard Buck, Hasenkamp
Hasenkamp 20
4950 Minden

SV d.Z. der Zwerg-Reichshuhner:
(German Reichhuhn bantams) Hans-Otto Jarecke
Ginsterweg 2
4513 Belm-Vehrte

SV d.Z. der Zwerg-Wyandotten: (Wyandotte Bantam)
Hant-Peter Flauaus
Marktstr. 32
6146 Alsbach-Hahnlein

SV d. Zuchter des Italierer-Huhnes: (Ancona Large)
Anton Muller
Spiegelslustweg 9
3550 Marburg

Holland

Anconaclub: B. Hanskamp
Steendijk 144
9404 AJ Assen

Antwerp Bearded Bantam: J. Krabshuis
Bolkhoeksweg 22
7607 TL Almelo

Assendelfter-N.H. Hoen: A. Goossens
Postbus 72
5324 ZH Ammerzoden

Austrlorpcl: Mrs. Ruyters
Kon. Wilhelminastraat 20
5321 TC Hedel

Barneveldercl: H. Westra
Verl. Stationstr. 29
9271 CG Zwaagwesteinde

Brabanter-Kraaikop-Uilebaardcl: A. Jansen
Parallelw. 120
6874 BH Wolfheze

Brahmacl.: C.V. Uythoven
Boeroestr. 50
2612 GG Delft

Japanese b.cl: C.V. IJsendijk
Imkerweg 65
1403 RA Bussum

Cochincl.: P. Verghoef
Wilh. Greevestr. 37
3172 VC Poortugaal

Drentse hoencl.: C. Oosterhoff
Hoofdweg 29
9341 BG Veenhuizen

Franse hoencl.: J. Tersteeg
Imminksplein 20
3402 BZ IJsselstein

Frisian cl.: S. Hoogstra
Buterhoek 51
9401 AA Berlikum

Domestic Waterfowlcl.: J.C. v.d. Zaan
Nespad 61
3474 KL Zegvel

Groningermewcl. W.D. Diemer
Bultsweg 273
7532 XE Glanerbrug

Nigerian Pair

Hollands hoen cl.: (Hamburgers) M. Aben
Beelenweg 15
5455 RH Wilbertsoord

Hollandse Krielcl.: (Dutch Bantams) G. Brouwer
Postbus 77
5360 AB Grave

Lakenveldercl.: G. Kreton
Julianastraat 21
3621 EG Breukelen

Langshancl.: J.G. Roelofs
H.v. Viandenstr. 12
8061 CW Hasselt

Ned. Hoend. Cl.: (All Dutch Breeds) W. Bleijenberg
Vossestaart 5
3648 HP Wilnis

Ned. Javacl.: (Rose Comb Bantams) Th. Wouterse
Stoofstr. 48
4693 RA Poortvliet

Ned. Kuif + baardk.cl.: (Polish) D. Diepenbroek
Zelhemseweg 29
7051 ZB Varsseveld

Leghorncl.: R. Woudwijk
Cort v.d. Lindenstr. 9
8802 RX Franeker

Minorcacl.: A. Hijma
Rijsstraatweg 22
9254 DJ Hardegarijp

Plymouthrockcl.: A.v.d. Brink
Zwolseweg 32
7315 GM Apeldoorn

Sabelpootcl.: (Booted Bantams) P. Verghaegen
Pr. Marijkestr. 8
6039 BX Stramproy

Sussex/Orpingtoncl.: (Dorkings) H. Berkers
Wolwever 3
5737 AD Lieshout

Wyandottecl.: A.A. Nieboer
Schoolstraat 37a
7681 CM Vroomshoop

New Hampshirecl.: W.v.d. Vleuten
Floreffestr. 107
5711 AC Someren

Rhode Island Redcl.: A Melissen
Heuningstr. 17
4501 GA Ochten

Sebrightcl.: C.v.d. Sijs
Ruysdaelstr. 98
2406 TH Alphen a/d Rijn

Twentse Hoencl.: A. Veldhuyzen
Stationsweg 106
3913 EV Woudenberg

Vechthoendercl.: (Game Fowl Bantams) W. Voskamp
Rijksweg 58
8121 EE Olst

Vorwerkhencl.: H. Wezendonk
Bleekveld 2
6913 AT Aerdt

Watermaalsebaardkr.cl.: H. Slijkhuis
Kerkstraat 80
6871 BM Renkum

Welsumerclub: P. Bouw
Brinkerweg 26
6721 ZC Bennekom

Pakistan Cock

Zijdehoenclub: (Silkies) E.H. Koot
Postbus 120
7670 AC Vriezenveen

Hungary

Hungarian Pigeon and Small Animals National Federation
Dembinszky u 52
Budapest 1071

Italy

Associazione Nazionale Allevatori Delle Specie Avicole
Sede Amministrativia: c/o APA (Bo)
Via Altabella 3
40126 Bologna

Unione Nazioale dell Avicultura: Ms. Rita Pasquarelli
Via Pasubio 4
00195 Rome

Cav Antonio Marazzini: S. Pellico
20053 Muggio
Milano

Ireland

Irish Society of Poultry Fanciers: R.T.B. Robinson
Rathbeggan
Dunshaughlin, County Meath

Japan

Nihon Aigankei Kyokai (Japan Poultry Association)
65 Ishigase, Oofu-town
Oofu-City
Aichi-Ken 474

Zen Nihon Chabo Hozon Kyokai:
(All Japan Bantam Association) Keiji Ogawa
4 - 10, Higashiazabu 3-chome
Minato-Ku, Tokyo 106

Saitama-Ken Chuo Keiyukai:
(Saitama-Ken Central Poultry Club) Kousaku Kashima
1205-1, Takasaka, Higashimatsuyama-shi
Saitama 355

Japan Wildlife Research Center: Book Center
39-12, Hongo 3-chome
Bunkyo-ku, Tokyo 113

Korea

Korea Poultry Association: Tae Won Chung
234, 5-Ka, Namdaemoon-ro
Chung-Ku, Seoul

Mexico

Mexico Poultry Club: Enrique Castellanos
Fco Madero No. 91
Tepatitlan, Jalico, C.P. 47600

New Zealand -North Island

North Island Poultry Pigeon & Cage Bird Association:
Mrs. B.Luff
P.O. Box 22
Private Bag
Waitotara, Wanganui

Auckland Poultry Poultry & Pigeon Association:
D. Walden
27 Rosella Road
Mangere East
Auckland

Hasting Poultry & Pigeon Association: R. Smith
703 Matai Street
Hastings

Levin Poultry & Pigeon Club: Mrs. B. Overall
17 Crowther Street
Shannon

Manawatu Poultry & Pigeon Club: G. Taylor
R.D. 7, Almadale Road
Feilding

Masterton Poultry & Pigeon Club: G.L. Wiblin
P.O. Box 104
Carterton

South Auckland Poultry & Pigeon Club: M. Keith
R. D. 4, Selman Road
Albany

Peru
Hen

Taranaki Agricultural Poultry & Pigeon Association:
G. Cadman
8 Rata Street
Inglewood

Waikato Poultry & Pigeon Club: M. Hanson
22 Irvine Street
Hamilton

Wanganui Poultry & Pigeon Fanciers Association:
G. Walker
Buxton Road
Wanganui

Wanganui Poultry & Pigeon Club: G. Belamy
R.D. 3, Old Kaiatea Road
Whangarei

Wellington Hutt Valley & District Poultry & Pigeon Club:
N. Hollebone
Takapu Road
Tawa
Wellington

New Plymouth Cage Bird Association: L. Robinson
11 Short Street
Waitara

New Zealand - South Island

South Island Poultry Pigeon & Cage Bird Association:
Ms. H. Thompson
Flat 2, 43 Peter Street
Ashburton

Invercargill Poultry Pigeon & Cage Bird Association:
K.A. Scott
7 R.D., Kennington
Invercargill

North Canterbury Poultry & Pigeon Club: R.T. Bryant
39 Queen Street
Rangiora

Dunedin Poultry Pigeon & Cage Bird Club:
Ms. D. Woodhead
Box 179
Mosgiel

Nelson Poultry & Pigeon Association: Ms. A.S. Fryer
67 Gladstone Road
Richmond, Nelson

Taieri Poultry Pigeon & Cage Bird Club: S. Cook
23 Fairview Terrace
Sawyers Bay, Otago

Ashburton Fanciers Society: B.J Glassey
37 Buckleys Terrace
Ashburton

Milton Utility Poultry & Bird Fanciers Club:
R. G. Robinson
21 George Street
Milton

Christ Church Poultry, Pigeon and Bird Club: R. Lamberts
Box 19
Lincoln

Waimate Poultry Club: Mrs. F. Woodham
8 R.D., Timaru Road
Waimate

Marlborough Poultry Pigeon & Cage Bird Club:
K. Sowman
8 North Street
Blenheim

Oamaru Poultry Pigeon & Canary Society: T.R. Hill
13 Reservoir Road
Oamaru

Norway

Norwegian Poultry Organization: Henrik Andreas Eddie
Berganvien 233
N-3135 Torod

Norsk Rasefjaerfeforbund Poultry Club: Ovin Uda
Skogen
4645 Nodeland

Scotland

Scots Grey Club: J. Robertson
21 Kersewell Avenue
Kaimend, Carnwath
Lanarkshire

South Africa

Bloemfontein Poultry Club: Charl Marais
P.O. Box 303
Bloemfontein 9300

Cape Town and Peninsula Poultry Club: Turner Haupt
24 Durbanville Avenue
Durbanville 7530

Natal and Coast Poultry Club: W.A. Stuart
P.O. Box 296
Merrivale 3291

Pretoria Poultry Club: D.G. Luies
P.O. Box 48795
Hercules 0030

Transvaal Game Fowl Club: F. Smit
146 Leyd Street
Rustenburg 0300

Vruburg Riethaan Poultry Club: Hannes Smit
11 Market Street
Vryburg 8600

West Rand Poultry Club: Ron MacHarry
P.O. Box 1322
Florida 1710

Phillipines Cock

Tanzania Cock

Spain

Associacion De Criadores De La Raza Prat
Carrer del Centre, 33
08820 El Prat Del Llobregat
Barcelona

Associacion De La Gallina Vasca
Eusko-oilozaleen Elkartea
Nafarroa Etorbidea, 6
20400 Tolosa (Guipuzcoa)

Associacion De Criadores De Aves El Francoli:
Jaime Huguet, 1
43800 Valls (Tarragona)

Associacion Espanola De Avicultura Artistica
Maldonado, 79
28006 Madrid

Sweden

Svensk Rasfjaderf Aforbundet Poultry Club
Gulli och Bror Persson
Hov Klagstrop
S-312 94 Laholm

Svensk Fagel Poultry Club
Box 26041
S-100 rl Stockholm

Swedish Poultry Club Association: Gosta Anderson
Ugglarp PL 8365
S-26200 Aengelholm

Switzerland

Schwizerischer Rassegelflugelverband Poultry Club: Ruedi Zingg
Ergeten
CH-9225 St. Pelagiberg

United States

American Bantam Association: Eleanor Vinhage
Box 127
Augusta, ME 07822

American Poultry Association: Karaen Porr
72 Springer Lane
New Cumberland, PA 17070-2455

American Langshan Club: Forest Beauford
Route 5, Box 75
Claremore, OK 74017

American Silkie Bantam Club: Mabel Witherall
2221 Blue Ridge Blvd.
Independence, MO 64052

American Wyandotte Breeders Association:
Stewart Jackson
6025 S. Eaton Lake
Littleton, CO 80123

Araucana Club of America: Leon Burch
Box 122
Littlefield, TX 79339

Chantecler Club of North America: Adrienne Blankenship
Box 43
Long Valley, NJ 07853

Dominque Club: Roger Voter
Route 2, Box 57A
Chadds Ford, PA 17364

National Frizzle Bantam Club: Dr. Loretta Schmidgall
Route 2, Box 785
Thomasville, PA 17364

National Jersey Giant Club: Meldra Harris
Box 683
Milledgeville, GA 31061

National Leghorn Club: Danny Jones
Route 3, Box 112
Middlepoint, OH 45863

Old English Game Bantam Club: Fred Jeffrey
Box 494
North Amherst, MA 01059

Oriental Fowl Fanciers Association: Horst Schmudde
209 Highway 34 North
Colts Neck, NJ 07722

Plymouth Rock Fanciers Club: Dave Adkins
Route 6, Box 310
Lucasville, OH 45648

Rhode Island Red Club: Emmett Rachels
Box 1161
Tavares, FL 32778

Silver Laced Wyandotte Club: Robert Coulter
Route 2, Box 151
Owatonna, MN 55060

SPPA - Society for the Preservation of Poultry Antiquities: Marion Nash
Box 102
Murphysboro, IL 62966

World Cochin Society: Bill Flitton
Route 1, Box 386
Collinsville, OK 74021

Selling boiled eggs with choice of sauces in West Africa

APPENDIX II
International Poultry Publications

Australia
Australian Poultry
P.O. Box 1908
Werribee, Victoria 3030

The Fanciers Gazette
291 Glaston Road
Glaston 2159
New South Wales

New Australian Poultry Journal
P.O. Box 1187
Hoffers Crossing 3030

Bangladesh
The Poultry Sangbad
110 Green Road (Farm Gate)
Ohaka

Canada
Feather Fancier
Route 5
Forest, Ontario NON IJO

Canada Poultryman
Suite 105A
9547 152
Surrey, British Columbia V3R 5Y5

China
China Poultry
46 Sangyuan Road
Yangzhou, Jiangsu Province 22500

Czechoslovkia
Chovatel Magazine
SZN, Vaclavski Namesti 47
1 - Nove Mesto
110 00 Prague

Denmark
Race Fjerkrcealv
Hilbert Jensen Publishing Company
Fenrisvej 6
7400 Herning

England

Fancy Fowl Magazine
Andover Road
Highclere
Newbury, Berks RG15 9PH

Feathered World
Winckley Publishing Company
37 Flyde Road
Preston, Lancs PR1 2XO

Poultry World International
18 Chapel Street
Petersfield, Hampshire GU32 3DA

World Poultry Science Journal
Cropthorne Cottage
Gravel Path
Berkamsted, Herts HP4 2PH

Finland

Siipkaraja Magazine
Keskuskatu 21
SF - 31600 Jokioinen

France

La Revue Avicole
19 Rue du Moulin
Abbecourt
02300 Chauny

Bantam Revuew
6 Rue Arisitide Briand Offemont
90300 Valdoie

Colombiculture
84 Rue Aristide Briand Offemont
90300 Valdoie

L_Aviculture
B.P. 6359
35963 Rennes, Cedex

Germany

Deutscher Kleintier Zuchter
Postfach 1642
7410 Reutlingen 1

Geflugel - Borse Magazine
Verlag Jurgens Publishing Company
Postfach 1529
8034 Germering

Deutscher Geflugel Ferbing
Reinharstrase 14
Postfach 130
1040 Berlin

Verlag Eugen Ulmenn
Nieb Uhrstrasse 53
5300 Bonn

Holland

Avi Cultura Magazine
Postbus 86
3958 Zv Amerongen
Netherlands

Fokkersbelangen Magazine
Postbus 210
5900 AE Venlo
Netherlands

World Poultry
Misset International
Box 4
7000 BA, Doetinchem
Holland

Hungary

Baromiipar (Poultry Industry)
Leninhorut 9 - 11
Budapest VII

India

Poultry Guide
20 Thyagaraj Nagar Market
Near Lodhi Colony
New Delhi 110 003

Thailand Pair
Fighting Chickens

Poultry Times
Sunder Villa
Ground Floor, Plot 717, 12th Road
Khar, Near Telephone Exchange
Bombay 400 052

Poultry Adviser
97 St. John's Church Road
Bangalore 560 005

Italy

Edaricole-Avicoltura Magazine
delis Calderini Stre,
32 Emilia Levantee
1-40139 Bologna

Israel

Meshek Haofot Magazine
P.O. Box 40005
Tel-Aviv 61400

Japan

Niwatori-No-Kenkyu Sha
623 Marunouchi Building
Chiyoda-Ku
Tokyo 100

Petlife Sha Company
Ikebuckuro Towa Bldg 4F
4-1-12 Ikebuckuro
Toshima-Ku
Tokyo 171

Ienohikari Kyoukai Magazine
11 Ichigaya Funagawara Chou
Schinjuka-Ku
Tokyo 162

Poultry Fanciers Newspaper
65 Ishigase, Ohbu-Machi
Ohbu-Shi
Aichi-Ken 474

Korea

Modern Poultry Magazine
6 Ga 283 (Sinsung B/D 302)
Dan San-Dong
Yeoung Deung Po-Gu
Seoul

Mexico

El Palenque de Oro
Apartado Postal No. 16-323
Delegaciion Azcapotzalco
CP 02000 Mexico City

New Zealand

New Zealand Poultry and Gamebirds
Box 5070
New Plymouth

Philippines

Agriscope
Basic Media Systems Company
Cityland VI
Sen. Gil Puyat Ave.
Makah, Metro Manila

Animal Husbandry & Agricultural Journal
Paencor, Inc.
Box 1295
154 Arenta Avenue
Quezon City

Russia

Ptitsevodstvo Magazine
V/O Agropromizdat
18 Sadovo - Spasskaya Street
Room 706
Moscow 1078907

South Africa

Poultry Gazette
c/o South African Poultry Show
Box 1202
Honeydew 2040

Pluimvee Poultry Magazine
Postbus 1192
Honeydew 2040

Spain

Selecciones Avicolas
Real Escuela de Avicultura
Plana del Paraiso 14
08350 Arenys de Mar
Barcelona

Sweden

Fjaderia
Svensk Fjaderiaskotsel
Langakravagan 56
S-241 35 Eslov

Svenska Tidskrift
Till, SFRS Ringforvaltare
Berit, Graab
Viredahol Sater
578 00 Aneby

Switzerland

Tieerwelt Magazine
Tieerwelt-Verlag
4800 Zofingen

Turkey

Animalia Poultry Magazine
Necla Cigerli
Sadicesme Sok. No. 11 Daire 2,
Senlik Apt.
Yusufpasa, Aksaray
Istanbul

Tavukcunun Sesi
Nuri Sozmen
Dolayoba Gozdagi Cad. No. 1, PK 3
Dolayoba
Istanbul

United States

Poultry Press
Box 542
Connersville, IN 47331

National Poultry News
Box 1647
Easley, SC 29641

The following publications are published by Watt Publishing Company, Mt. Morris, IL 61054

Broiler Industry
Egg Industry
Industria Avicola (Spanish)
Poultry Digest
Poultry International
Poultry International Chinese
Turkey World
Watts Poultry Statistical Yearbook, International Edition
Who's Who in the Egg and Poultry Industry
Who's Who in the International Egg and Poultry Industry

The Gamecock
Box 158
Hartford, AK 72938

Grit and Steel
Box 280
Gaffney, SC 29340

Yuglosavia

Peradarstvo Magazine
Vul. JNA
p. fah, 809
1101 Belgrade

Selling Chickens in Yemen Market

APPENDIX III
Poultry Shows

The largest and most popular poultry show in the United States is held in early November in Columbus, Ohio and is sponsored by Ohio State Poultry Breeders Association. For show dates, write: *Michael Stichler, President, 1120 SR 603, Greenwich, Ohio, 44837 or call: 419-895-1147.*

Though the U.S. holds many shows the following list was recommended by poultry fanciers. For additional information contact: William Wulff, Editor, Poultry Press, 317-827-0932.

West Coast Shows

Greater California Poultry Show
9344 North Fowler
Clovis, CA 93612

Pacific Coast Bantam Club Show
953 Encanada
Le Habra, CA 90631

Pacific Poultry Breeders Association Show
1933 Lovall Valley Court
Sonoma, CA 95476

Unique Device
for the grooming of chickens
for show

East Coast Shows

Northeastern Poultry Congress
30 Arthur Road
Weare, NH 03281

Boston Poultry Exposition
133 Millville Street
Mendon, MA 01756

Virginia Poultry Breeders Association
Route 3, Box 672
Ashland, VA 23005

For information regarding any poultry show held in the United States, write:
American Poultry Association,
72 Springer Lane, New Cumberland, PA, 17070
or call: 717-774-1926

The following is a list of leading international poultry shows.

Australia

Sydney Royal Easter Show
New South Wales

Melbourne Royal Show
Melbourne

Brisbane Exhibition
Brisbane

Dutch Poultry Show

Belgium

Europa-Expo
Gent

Brazil

Pequenos E. Medias
Sao Paulo

Denmark

Nyborg og Omegn fjerkreklub
Nyborg

England

National Championship Show
Royal Show Grounds
Warwickshire

National Federation Show
Stafford Show Grounds
Staffordshire

France

Exposition Internationale d'Aviculture de Mulhouse
Mulhouse

Exposition de Tours par la Societe Avicole de Touraine
Tours

Germany

Deutsch *Junggeflugelschau*
Hannover

National Rassegeflugel schaur
Nurenberg

Holland

Ornithophilia
Utrecht

South Africa

South African Championship Show
Royal Show Grounds
Pietermaritzburg

Switzerland

Schweizersche BSRG Rassegflugelausstellung
St. Gallen

APPENDIX IV
Exporters of Poultry

Ashdon Farms, Inc.
1251 Scugog Line 8
RR #1
Port Perry, Ontario
Canada L9L-1B2
Phone: 1-905-985-3555
Fax: 1-905-985-4005
Specializing in exportation of Poultry, Racing and Fancy Pigeons to all countries.

Australian Animals and Poultry, Ltd.
24 Ash Street
Sydney, New South Wales 2000
Australia

Roland Doer
12666 Brickyard Road
Shakopee, MN 55379

Domestic Fowl Trust
Honeybourne Pastures
Evesham, Worcestershire
WR11 50QJ England

Hainsworth International Livestock Limited
1 Peets Cottage
New Cut Lane,
Halsall, Ormskirk
L39 8SW England

Harold Halbach
Waterford, WI 53185

Heart of Missouri Poultry Farm
Box 954
Columbia, MS 65201

Clark Kidder
Milton, WI 53563

Rodney Michael
Box 244
Riverview, FL 33569

Miller Jersey Giant Farm
Bern, KS 66408

Morris Farms Hatchery
18370 S. W. 232nd Street
Goulds, FL 33170

McMurray Hatchery
Webster City, IA 50595

Hans L. Schippers
Goldschmedingplantsoen 15
De Hoenderhorst
1185 EM Amstelveen
Holland
Telephone/Fax: +31.20.6455227

Ed Shymanski
15 John Street
Old Bridge, NJ 08857

Suzuki Bird Farm
27-4 Honcho, 6 Chome
Shiki, Saitama 353
Japan

Duane Urch Poultry Farm
Route 1
Owatonna, MN 55060

Ralph Winter Guinea Farm
New Vienna, IA 52065

Painting
by
Mary Ellen Johnson

APPENDIX V

Recommended Reading

American Bantam Association, Color Bantam Standard
Available from:
Stromberg Chicks and Gamebirds Unlimited
Box 400
Pine River, MN 56474
USA

American Poultry Association, Color Poultry Standard
(All poultry, 464 color pictures)

Giavarini, Ida, Le Razze Dei Polli
(Chickens, excellent color pictures)
Available from:
Edagricole Publishing Company
via Emilia Levante
Postale 2157
40100 Bologna
Italy

Japanese Poultry Encyclopedia
(Chickens, exceptional color pictures)
Available from:
Noborou Suzuki
27 - 4 Honcho 6 - Chome
Shiki - shi
Saitama - ken
353 Japan

Orozco, Dr. Fernando, Razas De Gallines Espanolas
(Poultry of Spain, good color pictures)
Available from:
Ediciones Muni - Prense
Castello 37
280001 Madrid
Spain

Periquet, Jean Claude, Volailles De France
(Poultry breeds of France)
Available from:
Jean Claude Periquet
55400 Gincrey
France

Schmidt, Horst, Hanbuch der Nutz - und Rasselhuhner
(Chickens and Bantams, 300 color pictures)

Schmidt, Horst, Puten Perlhuhner Ganse Enten
(Geese, ducks, Turkeys, Guineas, 100 color pictures)
Available from:
Verllag Neumann - Neudem
Postfach 320
3508 Melsungen
Germany

Italian Painting

APPENDIX VI

Out of Print Books

Aldrovandi, Professor Ulisse, Aldrovandi on Chickens, 1600
Bement, C.N., American Poulterer_s Companion, New York, 1845
Brown, Sir Edward, Races of Domestic Poultry, 1906
Brown, Sir Edward, Poultry Breeding and Production (Set of three books) 1929, Caxton
Burnham, George, The History of Hen Fever, 1855, Boston
Cato, Varro, Columella, Palladius, Poultry Farming as Described by the Ancient Writers of Rome
De Reumur, M., The Art of Hatching & Bringing up Domestic Fowl
Dixon, Reverend Edmund, Ornamental and Domestic Poultry, 1848
Felch, I.K., Poultry Culture, 1887
Moubray, Bonnington, Moubray on Poultry
Smith, Page and Charles Daniel, The Chicken Book
Tegetmeier, W.B., The Poultry Book, 1853, London
Weir, Harrison, Our Poultry (Set of two books) 1902
Wright, Lewis, The Illustrated Book of Poultry, 1872, Cassell, Petter & Galpin

Antique Poultry Book Dealers

Pamela Fenwich
Box 263
Mayfield, New South Wales
2304 Australia

Silvio Mattacchione & Co.
1251 Scugog Line 8
RR #1
Port Perry, Ontario
Canada
L9L-1B2
Phone: 1-905-985-3555
Fax: 1-905-985-4005
Pays cash for one book or an entire library. Books of interest include poultry, waterfowl, peafowl, natural history, plants and pigeons. Send list with title and/or description and price.

Veronica Mayhew
Trewena, Behoes Lane
Woodcote, Near Redding
RG8 OPP, England

Bill Ziegler
10 Lincolnway West
New Oxford, PA 17350
USA

APPENDIX VII
Incubator Manufacturers

A.B. Incubators, Ltd.
40 Old Market Street,
Mendlesham
Suffolk
1I14 5SA
England

A. B. Incubators, Ltd.
1600 40th Street Ct.
Moline, IL
61265 USA

Brinsea Products Ltd.
Station Road
Sandford
Avon
BS19 5RA
England

Brower Incubators
P.O. Box 2000
Houghton, IA
52631 USA

Country Boy Incubator Specialist
3428 Beret Lane
Wheaton, MD
20906 USA

Curfew Incubators
Buttons Hill
Southminister Road
Althorne
Essex
CM3 6EN
England

Grumbach Brutegerate GmbH
Am Breitteil 2
Munchholzhausen
6330 Wetzlar 13
Germany

Humidaire Incubator Company
217 West Wayne Street
New Madison, OH
45346 USA

Kuhl Corportation
Kuhl Road
P.O. Box 26
Flemington, NJ
08822-0026 USA

Lyon Electric Company
765 Main Street
Chula Vista, CA
91907 USA

Miller Manufacturing Company Inc.
South Saint Paul, MN
USA

Patrick Pinker Ltd.
Latteridge
Iron Acton
Briston
BS17 1TY
England

Petersime Incubator Company
300 N. Bridge Street
Gettysburg, OH
45328 USA

Swan Creek Supply
12240 Spencer Road
Saginaw, MI
48609 USA

Werner Schumacher Ing.
Landenbach
Oberhessen
Germany

WHITE DORKING COCK.

PLYMOUTH ROCKS.

APPENDIX VIII

The National Poultry Museum and Heritage Center

The National Poultry Museum is dedicated to establishing a public museum and educational exhibition center which blends history, scientific data, recreational attributes and commercial production information. It tells the story of the American poultry industry and seeks to establish a library to preserve and present both antiquarian and modern publications and in doing so, broadens the understanding of and for American poultry. Your contributions are both needed and appreciated.

The National Poultry Museum and Heritage Center
630 Hall of Fame Drive
Bonners Springs, Kansas
66012 USA
Phone: 1-913-712-1075

By permission of the "Field"

VERY YOUNG LIGHT BRAHMA COCKEREL, 1853

APPENDIX IX

Other Books of Interest by Loyl Stromberg

ABC's of Poultry Genetics
A Guide to Better Hatching
Caponizing - Modern Management and Profitable Marketing
Dubbing Poultry and Why
Exhibiting Poultry for Pleasure & Profit
Guinea Fowl
Making Squab Raising Profitable
Peafowl Breeding & Management
Raising Game Birds in Captivity
Sexing - All Fowl
Successful Duck and Goose Raising
Swan Breeding & Management
Traps, Snares, and Making Cages

Stromberg's Chicks & Gamebirds Unlimited
Loy and Janet Stromberg
P.O. Box 400
Pine River, Mn. 56474

1-800-720-1134
1-218-587-2222
Fax: 1-218-587-4230

Painting
by
Mary Ellen Johnson

ADDITIONAL PUBLICATIONS

The Avian Embryo, #MM - 207
Cooperative Extension Service
Ohio State University
Columbus, OH
43210 USA

Bantam Standard
American Bantam Association
Box 464
Chicago, IL
60609 USA

Diseases of Poultry
Biester and Schwarte
Iowa State University Press
Ames, IA
50010 USA

From Eggs to Chick #878
College of Agriculture
University of Illinois
Urbana, IL
61801 USA

Manual of Poultry Diseases
Texas Agriculural Extension Service
Texas A & M University
College Station, TX
77844 USA

Merck Veterinary Manual
Merck & Comany
Rahway, NJ
07065 USA

Raising Poultry the Modern Way
Leonard S. Mercia
Garden Way Publishers
Charlotte, VT
05445 USA

Bibliography

Aldrovandi, Professor Ulisse, Aldrovandi on Chickens, 1600, translated from Latin, Dr. L.R.
Lind, 1963, University of Oklahoma Press
Aldrovandi, Professor Ulisse, Ornithologia, 1610, Frankfurt
Atkinson, H. The Life and Letters of John Harris the Corish Cocker, 1910, England
American Bantam Association, Book of Bantams, 1969
American Poultry Association, American Standard of Perfection, 1905
American Poultry Historical Society, American Poultry History, 1974
Belgium Poultry Standard, Trait D'Aviculture Sportive
Baldwin, J.P., Modern Game Bantams, 1936, England
Bantam Duck Standard, 1981, USA
Bement, C.N. American Poulterer's Companion, New York, 1845
Bennett, J.C., The Poultry Book, 1856, Boston
Brown, J.T. Encyclopedia of Poultry, 1909, Waverly
Brown, Sir Edward, Races of Domestic Poultry, 1906
Brown, Sir Edward, Poultry Breeding and Production, 1929, Caxton
Burnham, George, The China Fowl, Shanghae, Cochins, Brahmas, 1877
Burnham, George, History of Hen Fever, 1855, Boston
Campbell, E., The Orpingtons,
Cato, Varro, Columella, Palladius, Poultry Farming as Described by the Writer's of Ancient Rome
Choyselat, Prudent, The Discours Oeconomique, 1580
Croad, A.C., Langshan Fowls, 1877
Danish Breeding Standard

Darwin, Dr. Charles, The Variation of Animals and Plants Under Domestication, 1868
Delacour, J., The Waterfowl of the World, (Set of four books)
Delacour, J., The Pheasants of the World, 1977, England
De Reaumur, M., Art of Hatching and Bringing Up Domestic Fowl, 1750
Dickson, Walter B., Poultry: Their Breeding, Rearing, 1838
Dixon, Reverend Edmund, Ornamental and Domestic Poultry, 1848
Doyle, Martin, The Illustrarted Bookk of Domestic Poultry, 1854
Dryden, J., Poultry Breeding and Management, USA
Elington, W., Popular Poultry-Keeping for Amateurs,
Entwisle, W. F., Bantams, 1894, Wakefield
Finsterbusch, C.A., Cock Fighting all over the World, 1929, USA
Felch, I.K., The Breeding & Management of Poultry, 1877, USA
Felch, I.K., Poultry Culture, 1887
French Breeding Standard
German Breeding Standard
Gilbey, W., Sport in olden Time, 1912, London
Grit/Steel, Histories of Game Strains, 1929, USA
Hering, Leona, Poultry Firsts
Hicks, Stephan, Encyclopedia of Poultry (Set of two books)
Hodges, R.D., The Histology of the Fowl, 1974, London, New York, San Francisco
Howlett, R.H., The Royal Pastime of Cock Fighting, 1709, London
Howes, H., Turkeys & Geese, 1949
International Poultryman's Handbook, 1923, USA
Ives, Paul, Domestic Geese & Ducks, 1947, New York
Jeffrey, F., Bantam Chickens, 1974, USA
Johnson, Professor Willis, and George Brown, The Poultry Book, 1903
Johnstone, E., The AB.C. of Poultry, 1910
Jones, R., A Basic Poultry Guide, 1950
Lewis, M.M., The People's Practical Poultry Book, 1871, New York
Lombary, W., Hoenders Van De Wereld
May, C.G., American Breeds of Fowls, 1901
Megargee, Edwin, The Book of Poultry, 1926
Moubray, Bonnington, A Treatise on Domestic Poultry, Pigeons, and Rabbits, 1815
Mouniney, G.J., Poultry Products Technology, 1966, Westport

THE DORKING FOWL.*

National Geographic Magazine, April, 1927
Netherlands Breeding Standard
Newall, Venetia, An Egg at Easter
Norris, J. Books on Poultry and Cock Fighting, 1977, Paoli
Norris, L.C.R. and Elye, L., Brahmas and Cochins
Periquet, Jean Claude, Toutes Les Poules de France
Plant, William, Origin, Evolution, History and Distribution of the Domestic Fowl
Platt, Frank, All Breeds of Poultry, 1925
Proud, P., Bantams as a Hobby, 1901, London
Qiu, Professor Xiangpin, Poultry Breeds of China
Roberts, H. Armstrong, Commercial Poultry Raising, 1920, USA
Robinson, L., Modern Poultry Husbandry, 1953
Sauders, S.M., Domestic Poultry, 1875, New York
Schafer, Violet, Egg Craft
Schmidt, Horst, Manual of Utility and Fancy Chickens
Scott, G.R., The History of Cockfighting, 1960, London
Silk, W.H., Bantams & Miniature Fowl, 1951
Skelchley, W., The Cocker, 1814, London
Skinner, Professor John, American Poultry History, 1974
Smith, H.E., Bantams for Everyone, 1974, England
Smith, Page and Charles Daniel, The Chicken Book
Sonntag, Linda, Eggs
South African Breeding Standard
Switzerland Breeding Standard
Tegetmeier, W.B., The Poultry Book, London, 1853
Tegetmeier, W.B. and Harrison Weir, The Poultry Book, 1875
Thompson, Alan, The Complete Poultryman, 1952
Turner, G., Bantam Book, 1902, USA
Van Hoesen, Roy, History of Anconas, 1923
Weir, Harrison, Our Poultry, 1902
Wingfield, W.W. & Johnson, The Poultry Book, 1853, London
Wright, Lewis, The Illustrated Book of Poultry, 1872
Wright, Lewis, The Poultry Book, London, 1885

DORKING COCK